A Primer on Electric Vehicles in India

Saon Ray · Deb Mukherji
Editors

A Primer on Electric Vehicles in India

A Machine-Generated Literature Overview

Springer

Editors
Saon Ray
Indian Council for Research on
International Economic Relations (ICRIER)
New Delhi, Delhi, India

Deb Mukherji
ADM Prime Consulting Pvt. Ltd.
Gurugram, Haryana, India

ISBN 978-981-97-8965-8 ISBN 978-981-97-8966-5 (eBook)
https://doi.org/10.1007/978-981-97-8966-5

This Springer imprint is published by the registered company Springer Nature Singapore Pte Ltd.
The registered company address is: 152 Beach Road, #21-01/04 Gateway East, Singapore 189721, Singapore

Preface I

Auto-summaries can be generated by either an abstractive or extractive auto-summarization:

- An extraction-based summarizer identifies the most important sentences of a text and uses the original sentences to create the summary.
- An abstraction-based summarizer creates new text based on deep learning. New phrases are created to summarize the content.

The auto-summaries you will find in this book have been generated via an extractive summarization approach.

Each chapter was carefully edited by Saon Ray. The editors selected the papers which were then auto-summarized. The editors have not edited the auto-summaries due to the extraction-based approach and have not changed the original sentences. You will find the editors' reviews and guidance on the auto-summaries in their chapter introductions. In machine-generated books, editors are defined as those who curate the content for the book by selecting the papers to be auto-summarized and by organizing the output into a meaningful order. Next to the thoughtful curation of the papers, editors should guide the readers through the auto-summaries and make transparent why they selected the papers.

The ultimate goal is to provide a current literature review of Springer Nature publications on a given topic in order to support readers in overcoming information overload and to help them dive into a topic faster, to identify interdisciplinary overlaps, and to present papers which might not have been on the readers' radar.

Please note, that the selected papers are not used to train a LLM while the auto-summaries are created.

New Delhi, India — Saon Ray
Gurugram, India — Deb Mukherji

Preface II

This book tries to answer everything you want to know about electric vehicles in the Indian context. The book takes a value chain approach in its organization by discussing upstream and then the battery value chain. It then discusses the policies and regulations governing EVs and financing the EV transition. It discusses skills and jobs in the EV value chain. Through this book, we hope to answer some questions such as where is India in the EV transition. What are the major challenges in the EV ecosystem? What is working in the regulatory space? The book concludes with some thoughts on the way forward by bringing geopolitics into the discussion.

The key takeaways from the book are:

1. Understanding EVs—the big paradigm shift of the automotive industry.
2. Mapping out the scale of opportunities for Indian companies, both for domestic and export markets in the EV era.
3. List the supply chain challenges for battery raw material and technology gaps and how the Indian government and industry can work together to overcome these challenges.

This book will interest scholars and researchers in industrial economics, international trade studies, development economics, and business management as well as practitioners in energy transition and mobility transition, policymakers, government officials, and those in the corporate sector.

New Delhi, India — Saon Ray
Gurugram, India — Deb Mukherji

Contents

About the Editors

Prof. Saon Ray is an economist specializing in international trade and industry issues with over 24 years of experience. Her areas of interest include global value chains, technological upgrading of Indian industries, free trade agreements and trade creation effects, technology transfer, foreign direct investment, efficiency and productivity of firms, financial inclusion, energy, and climate change related issues. Her Ph.D. in Economics from Jawaharlal Nehru University examined the role of intellectual property rights in transferring technology to developing countries. Her latest book from Springer is *Global Cooperation and G20 Role of Finance Track*.

Dr. Deb Mukherji is a professional from the Indian Automotive industry with more than 35 years of working experience. He has been CEO of major multinational companies in India from Japan, Korea, and Europe. A mechanical engineer by profession, he did his Ph.D. from Jamia Millia Islamia with research work on assessing global competitiveness of Indian automotive components' industry comparing with Chinese industry. He has been actively involved in automotive industry transition from IC to Electric and successfully set up an electric vehicle manufacturing company with local manufacturing of all critical parts in India. He speaks on various academic and industry forums guiding industry professionals and students on the opportunities and challenges in the upcoming Electric Vehicles Industry.

Part I
Electric Vehicles

Part I has a chapter to introduce the book and its organization. The book takes a value chain approach in its organization by discussing upstream and then the battery value chain. It then discusses the policies and regulations governing EVs, and financing the EV transition. It also discusses skills and jobs in the EV value chain.

Chapter 1
Introduction

1.1 Introduction by Editor

1.1.1 Introduction

At COP 27, India announced its Long Term Low Emission Development Strategy (LT-LEDS) and committed to reducing its net emissions to zero by 2070.[1] Decarbonizing transport can be an important way in which this can be attained in India.[2] Electric vehicles (EVs) are important in reducing emissions.[3] The life cycle assessment (LCA) shows that the largest reduction in GHG emissions can be achieved by urban passenger transport through the shift of cars and 2-wheelers to electric buses.[4] Throughout the world, countries are transitioning to EVs. This book looks at the Indian EV sector. What would be EV transition look like for India? The book takes a value chain approach in its organization by discussing upstream and then the battery value chain. At the same time, the EV transition could also be an opportunity for the Indian auto sector.[5] However, there are many challenges that remain. These are also discussed.

[1] MOEFCC (2022). At COP 27, India updated its Nationally Determined Contribution (NDC) under the Paris Agreement which will lead to net-zero emissions by 2070 (Nandi, 2022).

[2] Kathuria et al. (2018). Aziz (2018).

[3] Muncrief (2021).

[4] ITF (2023).

[5] See Wadhwa (2023). However, will this extend to the EV sector also?

S. Ray and D. Mukherji (eds.), *A Primer on Electric Vehicles in India*,
https://doi.org/10.1007/978-981-97-8966-5_1

1.1.2 India and EV Transition

The uniqueness of the EV transition in India is that it is driven by 2 and 3-wheelers. According to data from VAHAAN, the total number of electric vehicles is 3850267 of which 2026471 is 2-wheelers and 1648860 3-wheelers (SMEV). The impetus for the EV transition in the country has been driven by the policies undertaken by the Government of India. In 2019, the National Mission on Transformative Mobility and Battery Storage was approved by the Cabinet, and the Phased Manufacturing Programme was announced.[6] By 2030, it is expected that 30% of private cars (4-wheelers), 70% of commercial vehicles, and 80% of 2 and 3-wheelers will be electric.[7] What are the challenges in meeting this target?[8] We discuss many challenges encountered in meeting the challenge.[9] On the other hand, Tata has emerged as the only Indian Original Equipment Manufacturer (OEM) (among 20 largest in the world) selling EVs.[10]

1.1.3 Organisation of the Book

The book is organized into ten chapters and four parts.

Part I: Electric Vehicles

Part I has a chapter to introduce the book.

Chapter 1: Introduction sets the tone for the book and motivates it. It briefly discusses the statistics on EVs in India in terms of assembly/manufacturing, and uptake.

Part II: Upstream

Part II contains Chaps. 2 and 3. These chapters discuss the upstream value chain of the battery chain and focus on the minerals and technology used to produce EV batteries.

Chapter 2: Sourcing Minerals for the Battery

The battery of the EV uses various minerals like lithium and cobalt. Lithium is a key component of batteries and other electronics. The major use of cobalt is in chemicals

---

[6] https://pib.gov.in/PressReleaseIframePage.aspx?PRID=1567807

[7] Arthur D. Little report.

[8] https://www.downtoearth.org.in/blog/energy/how-realistic-are-india-S-2030-fleet-electrification-targets--88853#:~:text=By%202030%2C%20the%20government%20intends,MoRTH)%2C%20stated%20in%202021.

[9] We do not discuss the issue from the electric mobility or the urbanization angle. There are many articles on this aspect. See: https://theicct.org/publication/hdv-india-delhi-ev-policy-jun23/

[10] https://cleanmobilityshift.com/industry/tata-the-only-indian-oem-to-feature-in-the-worlds-top-20-automakers-that-sell-evs/

(paints and dyes) and corrosion-resistant alloys. Cobalt is difficult to substitute—it has some properties that make EV batteries stable. In EVs, the battery accounts for 70% of the cost of two-wheelers and 50% of cars. All EV production depends on the supply of cobalt and lithium from a few countries. 65% of lithium reserves are in Bolivia and Chile, while 60% of cobalt reserves are in Congo. For all countries, securing a supply of essential metals is critical. This chapter will discuss the various aspects of lithium and cobalt and other critical minerals sources for EV batteries in India.

Chapter 3: Technology

The nature of the technology of the batteries will determine the material that is used. In lithium-ion batteries, lithium is used predominantly. However, R&D is going on worldwide (partly to offset the scarcity of lithium). In the future, some of these technologies will become commercial and will determine the demand for critical minerals in future. How should a country like India decide its strategy both towards technology acquisition and sourcing minerals to produce low cost EVs?

Part II: Making in India

This section on Manufacturing/Assembling in India discusses the EV value chain in the context of India. This part has three chapters. Chapter 4 is on the value chain. Chapter 5 is on the imports of raw materials like lithium and components like batteries which are not manufactured in the country. Chapter 6 discusses the future of the industry in terms of India's attracting FDI and its export potential.

Chapter 4: Value Chain for Manufacture and Assembly

The value chain of the EV for manufacture or assembly comprises of several steps. Certain parts are common to both ICE and EV: e.g. clutch/torque control, start/stop system, pistons, spark plugs, and turbochargers. The new parts include a power distribution module, onboard charger, battery management system, vehicle interface control module, DC/DC converter, etc. The most critical part is the battery. Every battery has two electrodes —a cathode and an anode—between which trillions of charged lithium atoms travel (in case of the lithium-ion batteries). The cathode largely determines a battery's performance, cost and environmental footprint. Cathode today is produced almost entirely in Asia. This chapter will look at India's status in assembly and future manufacturing of batteries.

Chapter 5: Imports and Exports of Components

Currently, certain components are imported. The chapter answers some questions like: what are those components? How can India develop capability in these components? Is the PLI scheme adequate to address this challenge to develop capability? This question will be answered in Chap. 7.

Besides components, raw materials are also imported. Raw materials like lithium and cobalt are imported. Where are they imported from? Are there alternative sources for these materials?

Chapter 6: FDI in the EV Sector

The EV transition in India will require considerable investments including FDI. The chapter will document the nature and volume of FDI flowing into the sector. FDI brings along technology with it. Some of the other questions that this chapter is asking are what are the barriers to investment? Is technology transfer also taking place? Will this help India develop its capability in the sector?

Part III: Ecosystem

Part III will discuss the evolution of the ecosystem in India. It will discuss the conditions that lead to the development of an ecosystem and how an ecosystem evolves thereafter. It will also discuss the role of industrial and trade policy in this context. It will discuss these issues in Chap. 7 on policies and regulations in the EV space. Chapter 8 is on financing the EV transition in India. Finally, Chap. 9 discusses skills and employment in the EV industry.

Chapter 7: Policies and Regulations at the Central Level and States

At the centre, there are policies like FAME I and FAME II.[11] PLI scheme for Advanced Cell Chemistry has also been announced. In addition, most states have an EV policy. Are these policies and regulations enough to usher in an EV transition?[12] Urban bodies have a role to play in the adoption of EVs.[13] What can India do to attract more investment? Are there gaps that must be overcome? Are these policies conducive to developing the EV ecosystem? These are some of the issues that this chapter will address.

Chapter 8: Financing EV Transition in India

Financing the EV transition will be key. This chapter will examine various ways this financing is being done and capital is being raised. It will discuss if private domestic investment is flowing into the sector. It will discuss whether the environment is conducive to FDI and whether the return is commensurate.

Chapter 9: Skills and Workforce Training

The transition from internal combustion engine (ICE) to EVs will require changes in the workforce. Certain skills are required in ICE which may become redundant in EVs. Certain parts are common to both ICE and EV: e.g. clutch/torque control, start/stop system, pistons, spark plugs, and turbochargers. The new parts include a power distribution module, onboard charger, battery management system, vehicle interface control module, DC/DC converter, etc. The most critical part is the battery.

---

[11] The subsidy on 2-wheelers has been cut recently. See Baruah (2023).

[12] Recently, there have been some concerns regarding the withdrawal of subsidies paid to firms that have not met the localization requirements. https://www.livemint.com/opinion/online-views/our-ev-subsidy-policy-is-due-for-a-full-overhaul-11694531840905.html

[13] Urban Local Bodies have a key role in accelerating EV adoption in cities. We do not discuss this issue in the book. Link: https://cleanmobilityshift.com/policy-regulation/urban-local-bodies-have-a-key-role-in-accelerating-in-ev-adoption-in-cities/

This chapter will discuss the skills required in EV assembly, manufacturing, and repair. It will examine the training necessary for inculcating this skill set in workers. The chapter will also discuss what the EV transition will mean for the country in terms of the workforce engaged in ICE versus EVs.

Part IV: Looking Ahead

The final part of the book examines the way ahead, the geopolitics behind mineral sourcing, and the path for EV transition.

Chapter 10: Conclusion and Way Ahead

The final chapter will bring together all the strands explored in the book by way of a conclusion. It will discuss some possible pathways for the EV transition in India. The pathways will depend on a host of factors including geopolitics.

1.1.4 USP of the Book

This book tries to answer everything you wanted to know about electric vehicles in the Indian context.

The key takeaways from the book are:

1. Understanding EVs - the big paradigm shift of the automotive industry.
2. Mapping out the scale of opportunities for Indian companies, both for domestic and export markets in the EV era.
3. Listing the supply chain challenges for battery raw material and technology gaps and how the Indian government and industry can work together to overcome these challenges.

The USP of the book is that it is a human–machine collaboration.

Machine Generated

Sun et al. (2022) study the relationship between economic growth, CO_2 emissions, and battery electric vehicles in five countries (the United States of America (USA), China, France, Germany, and Norway) using panel data from 2010 to 2020. Gupta and Garg (2022) examine the enablers for EV adoption in India and observe that Government Policies (incentives include granting subsidies, reduction in taxes, relaxation in paying the fees associated with owning a vehicle, etc.) and initiatives and Technological Development are two main enablers. Rehman and Seth (2023) examine the enablers for EV penetration in India in the context of urban sprawl. Das and Ghosh (2023) develop a framework for achieving the vision of net zero for India using bibliometric analysis. Ammaiyappan et al. (2019) examine the reliability of an EV based on the thermal analysis of the battery and power converter, and also the effects of losses related to driving behavior. Kumar et al. (2022) study the number of constraints and available time slots for charging/discharging in the Indian context, which differ with region. Arora et al. (2022) study consumers' purchase intentions toward EVs in the Indian market characterized by poor awareness and a slow diffusion process.

References

Arthur D. Little (2022) Unlocking India's electric mobility potential. Report. Link: https://www.adlittle.com/en/insights/report/unlocking-india%E2%80%99s-electric-mobility-potential

Aziz, Z. (2018) Low-Carbon Pathways for Urban Development and Mobility in India. In Kathuria, R., Ray, S., and Bandyopadhyay, K. (2018) (eds.) *Low Carbon Pathways for Growth in India*, Springer. https://link.springer.com/book/https://doi.org/10.1007/978-981-13-0905-2.

Baruah, R. (2023) Govt cuts subsidy on electric 2-wheelers. Livemint. May 21. Link: https://www.livemint.com/companies/news/govt-cuts-subsidy-on-electric-2wheelers-11684690738134.html

Bhattacharjee, A. (2023) Tata—the only Indian OEM to feature in the world's top 20 automakers that sell EVs. Clean Mobility Shift. June 13. Link: https://cleanmobilityshift.com/industry/tata-the-only-indian-oem-to-feature-in-the-worlds-top-20-automakers-that-sell-evs/

ITF (2023) Life-cycle Assessment of Passenger Transport: An Indian Case Study. Report. August. Link: https://www.itf-oecd.org/life-cycle-assessment-passenger-transport-indian-case-study-launch-event#:~:text=The%20study%20uses%20the%20ITF,%25%20renewable%20energy%20%E2%80%93%20is%20needed.

Kathuria, R., Ray, S., and Bandyopadhyay, K. (2018) (eds.) *Low Carbon Pathways for Growth in India*, Springer. https://link.springer.com/book/https://doi.org/10.1007/978-981-13-0905-2.

Kaur, H., and Narla, A. (2023) Support Battery Electric Trucks in the Delhi EV Policy 2.0. ICCT Position Brief. Link: https://theicct.org/publication/hdv-india-delhi-ev-policy-jun23/

MOEFCC (2022) India Submits its Long-Term Low Emission Development Strategy to UNFCCC. PIB. November 14. Link: https://pib.gov.in/PressReleasePage.aspx?PRID=1875816

Muncrief, R. (2021) Why are Electric Vehicles the Only Way to Quickly and Substantially Decarbonize Transport? ICCT Blog. Link: https://theicct.org/why-are-electric-vehicles-the-only-way-to-quickly-and-substantially-decarbonize-transport/

Nandi, J. (2022) COP27: India unveils strategy for net zero goal by 2070, takes moral high ground. Hindustan Times. November 14. Link: https://www.hindustantimes.com/india-news/cop27-india-unveils-strategy-for-net-zero-goal-by-2070-takes-moral-high-ground-101668447611871.html

PIB (2019) Mobility Solutions get a boost National Mission on Transformative Mobility and Battery Storage approved by Cabinet. March 7. Link: https://pib.gov.in/PressReleaseIframePage.aspx?PRID=1567807

SMEV (n.d.) EV Industry Electric Vehicle Sales Dashboard. Link: https://www.smev.in/statistics

Tripathi, M. (2023) How realistic are India's 2030 fleet electrification targets? Down to Earth. Link: https://www.downtoearth.org.in/blog/energy/how-realistic-are-india-s-2030-fleet-electrification-targets--88853#:~:text=By%202030%2C%20the%20government%20intends,MoRTH)%2C%20stated%20in%202021

Wadhwa, N. (2023) Indian automotive industry poised to become the next China says ZF. Financial Express. May 30. Link: https://www.financialexpress.com/business/express-mobility-indian-automotive-industry-poised-to-become-the-next-china-says-zf-3107819/

1.2 Machine Generated Summaries

Disclaimer: The summaries in this chapter were generated from Springer Nature publications using extractive AI auto-summarization: An extraction-based summarizer aims to identify the most important sentences of a text using an algorithm and uses those original sentences to create the auto-summary (unlike generative AI). As the constituted sentences are machine selected, they may not fully reflect the body of the work, so we strongly advise that the original content is read and cited. The auto generated summaries were curated by the editor to meet Springer Nature publication standards. To cite this content, please refer to the original papers.

Machine generated keywords: relationship, cop, population, scheduling, reliability, enabler, decarbonisation, thermal, simulation, discharge, ancillary, transportation system, acid, co, conventional fuel.

1.2.1 An Investigation on the Role of Electric Vehicles in Alleviating Environmental Pollution: Evidence from Five Leading Economies [1]

This is a machine-generated summary of:

Sun, Dongying; Kyere, Francis; Sampene, Agyemang Kwasi; Asante, Dennis; Kumah, Naana Yaa Gyamea: An investigation on the role of electric vehicles in alleviating environmental pollution: evidence from five leading economies [1].

Published in: Environmental Science and Pollution Research (2022).

Link to original: https://doi.org/10.1007/s11356-022-23386-x.

If you want to cite the papers, please refer to the original.

For technical reasons we could not place the page where the original quote is coming from.

Abstract-Summary

"The relationship between battery electric vehicles (BEV) and carbon dioxide emission (CO_2) has significant environmental outcomes."

"Notwithstanding, battery electric vehicles have not been extensively explored through econometric approach."

"For countries to meet their net zero targets, it is crucial to consider the role of battery electric vehicles, renewable energy consumption, and CO_2."

"This study therefore examines the dynamic correlation between BEV, gross domestic product (GDP), urbanization (URB), renewable energy consumption (REC), population (POP), and CO_2 in five leading countries (the United States of America (USA), China, France, Germany, and Norway) using panel data from 2010 to 2020."

"The study applied the common correlated effect mean group (CCEMG) and augmented mean group (AMG) approach to ascertain the robustness of the long-run relationships among the variables."

"BEVs were found to be more energy efficient and the adoption of renewable energy through the manufacturing and battery production process would reduce CO_2 emission especially in China and the USA."

"The research proposed several policy implications for policy and decision-makers in the five leading countries for combating climate change and increasing productivity in the electric vehicle market and renewable energy consumption."

Introduction

"The implementation of these models in the current study, for instance, will guide the assessment of the correlation between carbon emissions, BEV adoption, GDP, urbanization, renewable energy demand, and population in this study."

"The literature will be discussed in this section exploring the linkage between carbon emissions (CO_2), battery electric vehicles (BEVs), GDP, population growth, urbanization, and renewable energy consumption."

"Several investigations have ascertained the influence of electric vehicles on carbon emissions in recent years; however, very few uses econometric approaches to examine the relationship between electric vehicles (BEV) and carbon emissions, particularly in the USA, China, Germany, France, and Norway."

"No studies have examined how CO_2 emissions engage the five variables: BEV, urbanization, GDP, population growth, and renewable energy consumption, through an empirical investigation using econometric series data from the sampled countries between 2010 and 2020."

Empirical Methodology

"Empirical studies that use the Kuznets environmental curve assertions find a significant influence on economic growth and carbon emissions."

"This research provides evidence for renewable energy consumption (REC) critical role in reducing carbon emissions and boosting economic growth."

"The researchers of this study anticipate a positive linkage between REC and carbon emissions in the sample countries."

"Using a panel data and macroeconomic model to estimate levels of CO_2 in the EU from 2000 to 2019, Busu and Nedelcu [2] and Sampene and others [3] conclude that urbanization positively affects CO_2 emissions; thus, this study asserts that URB will have a positive interaction with CO_2."

"The approximate variables used in this study are economic growth (measured as GPD constant 2010 US$), total energy generated from renewable sources (calculated as million kilowatt-hours), and CO_2 emissions (quantified in units of metric tons equivalent per capita)."

Panel Cointegration Test

"A panel cointegration method is employed to calculate the long-run equilibrium linkage between two or more variables when the variables have a unit root."

"The study employed the Westerlund [4] cointegration technique to explore the series' cross-sectional dependency (CSD) and heterogeneity."

Panel Estimation Approach

"The study employed the CS-ARDL introduced by Chudik and Pesaran [5] in the investigation to ascertain the short- and long-run projections."

"The CS-ARDL approach aid the resolution of heteroscedasticity, endogeneity, and linear relationship between the variables (Baydoun and Aga [6]; Huang and others [7])."

"Chudik and Pesaran [5] argued that the ARDL model is one of the most "heterogeneous panel data estimators"."

"The methods entitled as CS-ARDL are feasible to apply whenever there is a presence of heterogeneity and the CSD among the variables of interest. Chudik and Pesaran [5] argued that CS-ARDL model basically grounds the ARDL model with the element of linear combination of average CSD."

Robustness Test

"The AMG and CCEMG models proposed by Teal and Eberhardt [8] and Pesaran [9] were also adopted in our research to ascertain the robustness of the short- and long-run estimation methods."

"The AMG and CCEMG modeling techniques are utilized as they are robust and produce unbiased and accurate estimates."

Panel Causality Testing

"The study applied the Dumitrescu and Hurlin [10] to ascertain the causality test among the variables under consideration, i.e., CO_2, urbanization, GDP, BEV, renewable energy consumption, and population."

"The non-causality test for Dumitrescu and Hurlin [10] considers the heterogeneous panels and CSD."

"In line with the panel causation test by Ahmad and others [11], the Dumitrescu and Hurlin was used to improve the causality diagnosis in this investigation."

Empirical Results and Discussions

"The authors conclude that the sampled countries development have a relationship with GDP, CO_2, electric vehicles, urbanization, population, and renewable energy consumption, premised on these intriguing findings."

"Adoption of BEVs, for instance, can result in significant savings in countries such Norway, where renewable energy makes for a significant share of the power mix."

"Regarding renewable energy consumption, the CS-ARDL estimator shows that this variable adversely influences CO_2 emissions as the study countries depend on coal and oil."

"Manufacturing companies in industrialized or semi-industrialized countries are more appealing and efficient in their production of goods and services and spur economic growth; however, this activity accelerates the use of non-renewable source of energy, thereby increasing pollution levels."

"The study countries with renewable energy benefits and advantages should pursue a heterogeneous policy of increasing investment in BEV infrastructures and broadening the renewable energy resource base."

Conclusion and Policy Recommendations

"Using annual data from five leading countries (USA, China, France, Germany, and Norway), this research explored the dynamic relationship between economic growth, battery electric vehicles, urbanization, renewable energy consumption, population, and CO_2."

"This study summarized the findings as follows: (1) the study discovered a significant positive correlation between economic growth, population, and CO_2. (2) Battery electric vehicles, renewable energy consumption, and carbon dioxide emissions were discovered to have a negative relationship. (3) The panel D-H causality test demonstrates a unidirectional causal relationship that runs from economic growth, battery electric vehicles, and urbanization to carbon dioxide emissions. (4) The findings of D-H causality reveal a two-way causality among renewable energy consumption, population, and CO_2 emissions."

"BEVs may be thought of as battery storage for stabilizing electric networks supplied by fluctuating renewable sources of energy, resulting in more sustainable power use."

1.2.2 Modelling the Enablers for Adoption of Electric Vehicles in India [12]

This is a machine-generated summary of:

Gupta, Anshu; Garg, Ananya: Modelling the enablers for adoption of electric vehicles in India [12].

Published in: International Journal of System Assurance Engineering and Management (2022).

Link to original: https://doi.org/10.1007/s13198-022-01687-1.

If you want to cite the papers, please refer to the original.

For technical reasons we could not place the page where the original quote is coming from.

Abstract-Summary

"Electric vehicles in this context are recognised as a promising green alternative to traditional gasoline vehicles, for reducing the dependency on crude oil, consumption of energy and emissions of greenhouse gases from the transport sector."

"Talking about countries like India, currently electric vehicles comprise less than 1% of the overall automotive market share."

"The Indian Government is encouraging the adoption of electric vehicles through several schemes and policies but the results are yet to be seen."

"The purpose of this study is to analyse the relationships between the major drivers that would support the adoption of the electric vehicles in India using Interpretive Structural Modeling Technique."

"The results unveil that the Government Policies and Initiatives and the Technological Advancements could be the key drivers in accelerating the adoption process."

Introduction

"Industrial revolutions have bestowed humans with better infrastructure, higher standard of life, wealth and access to resources, motorised transportation and a globalised world."

"Post industrial revolution the world's population has grown tremendously especially in the developing countries like India."

"Amplification of transportation activities accounts for a significant impact on the environment which call for development and adoption of sustainable transportation alternatives."

"Electric vehicles in this context are recognised as a promising green alternative to traditional gasoline vehicles, for reducing the dependency on crude oil, consumption of energy and emissions of greenhouse gases from the transport sector (IEA [13]; Eberle and Helmolt [14])."

"It aims to develop India as the largest manufacturer and R&D hub for the automobile industry across the world."

"Studies in the literature have discussed how the momentum of electrification has been escalated in various countries through policy initiatives, infrastructural development and collaborations with various stakeholders (WRI India [15])."

Literature Review

"Krause and others [16] highlighted the measures that are needed to be undertaken in order to disseminate the required information to the consumers related to the public incentives available and the superior benefits of EV."

"These factors include financial, technical, infrastructural and policy enablers which if set into the place, will drive the adoption process."

"Infrastructural Development Lieven [6] suggested that low level of infrastructure related to the charging facilities is a major concern and is a cause for non-adoption of EVs amongst the consumers."

"While applying the brakes, the vehicle loses a lot of energy, which is transferred to the battery of the EVs through regenerative braking technology (Namdeo and others [17]; Joshi and others [18])."

"Longer range of vehicles along with better operational characteristics will provide a boost to the EV segment and will increase the likelihood of buying (Liao and others [19])."

Research Gap and Contribution

"Literature has been identified related to the demand forecasting effect that EVs may have on the environment if the technology is adopted, but there are a few studies talking about the key enablers that, if customized according to the current situations could help in attaining the objectives."

"Since India is a country which is still at the inception stage of adopting EVs, the various enablers need to be identified to ease the penetration process."

"The study hence aims to identify the major enablers that would facilitate the adoption process in India."

"The enablers for adoption of EVs are identified through detailed review of literature and in discussion with experts."

"The study also uses the MICMAC analysis to further analyse enablers on the basis of their driving and the dependency power."

"This provides a better picture in front of the policymakers or the decision makers to study the mutual relationships and dependency lying between the enablers."

Research Methodology

"Step 2 To study the relationship between the enablers a Structural Self-Interaction Matrix (SSIM) is developed."

"Step 6 Lastly a digraph indicating the relationships between the enablers is obtained."

"V for the relation E_i to E_j signifying that enabler E_i will help attaining enabler E_j."

"A for the relation E_j to E_i signifying that enabler E_j will help attaining enabler E_i."

"O to indicate no relation between the enablers where E_i and E_j."

"Cross-impact matrix multiplication applied to the classification analysis is conducted in order to study the relationship between the enablers through the driving and the dependence powers."

"All the enablers are plotted according to their driving and dependency powers on the graph."

"The first cluster represent the autonomous enablers which shows weak dependence and driving power."

"The second cluster represent the dependent enablers that shows high dependency but low driving power."

Results and Discussion

"This study aimed to investigate the effectiveness of enablers that are required to drive the implementation of electric vehicle plans in India as an initiative towards sustainable transportation."

"The results show that Government Policies and Initiatives and Technological Development are two main enablers lying in the fourth quadrant that demand an immediate action to drive the change."

"These incentives include granting subsidies, reduction in taxes, relaxation in paying the fees associated with owning a vehicle etc. The companies and the manufacturers could also get a boost and be motivated to start producing more only if the government deploys adequate charging infrastructures facilities, indulge them into the partnerships with the private companies to provide adequate funding, resources for research and development and extend financial incentives."

"Technological Development holding a strong driving power also plays an important role which not only reduces the cost of the vehicle but also makes the country capable enough to produce locally."

Conclusion

"In order to find the enablers which would support the introduction of EVs in India a detailed literature review was conducted."

"The prime concern of the study is to examine the relationship between the enablers."

"We could gauge from the findings that India is still not ready for the EV revolution as they need to set in place many policies and should start with deploying resources in the cities having basic infrastructure already set in place."

"Model derived in the study has wide application, could be used for improving the performance, achieving the efficiency and effectiveness of operations and making strategy decisions for policy formulations."

"Once these enablers are set into place in India, one can believe that this sustainable solution is viable."

"To reach the exact conclusions and analysis future research can focus on quantifying the enablers."

1.2.3 Investigation and Modeling of Electric Vehicle Enablers (EVE) for Successful Penetration in Context to India: Mitigating the Effect of Urban Sprawl on Transportation [20]

This is a machine-generated summary of:

Rehman, Minhaj Ahemad; Seth, Dinesh: Investigation and modeling of electric vehicle enablers (EVE) for successful penetration in context to India: mitigating the effect of urban sprawl on transportation [20].

Published in: Environmental Science and Pollution Research (2023).

Link to original: https://doi.org/10.1007/s11356-023-26022-4.

If you want to cite the papers, please refer to the original.

For technical reasons we could not place the page where the original quote is coming from.

Abstract-Summary

"Government at both central and state levels is dealing with this challenge in two ways-adding more infrastructure for public transport and encouraging electrical vehicles (EVs)."

"Adoption of EVs for public mobility is eco-friendlier and economic."

"It is observed that EV penetration in many pockets is not growing and is yet to mature for usage."

"Research mainly focused on reporting barriers and did not guide EV penetration enablers."

"This study bridges the research gap and offers useful insights about EV penetration phenomenon and makes use of both qualitative and quantitative treatments."

"The study reports six key enablers, which are-developing high-capacity batteries with short recharge time, improving service support, framing promotive government policies, lowering electricity tariffs using sustainable and reliable sources, and reducing dependence on imported raw materials."

"These enablers need an urgent attention from the industries and researchers for successful EV penetration in Indian context."

Introduction

"As EVs have become a preferred and viable choice over conventional fuel-based road-vehicles."

"Worldwide concerns about automobile emissions, because of conventional fuels-based vehicles for road transport, have been growing."

"As carbon dioxide emissions along with natural resources depletion are pressurizing economies around the world to deploy EV-based solutions to replace conventional fuel-based vehicles and expedite the transition to a low-carbon transportation system while having no or minimal negative environmental impacts (Tarei and others [21]; Patyal and others [22])."

"This study not only attempts identification of enablers but it also models enablers to understand relationships, positioning influencing managerial decisions and then attempts validation. (1) To identify penetration responsible EV enablers (EVEs) in Indian context through literature review (2) To model identified EV enablers to establish inter-relationships using ISM and to validate with BWM method (3) To prioritize and rank enablers based on their positioning and relative importance This study is organized as follows."

Review of Literature and Identification of EV Enablers

"According to (Goel and others [23]), EV penetration in India may be influenced by variables like technology development, reduced vehicle costs, supportive government policies, incentives for owners, parking benefits, and good public charging infrastructure facilities."

"According to (Tarei and others [21]), important factors influencing EV penetration are number and density of charging stations and infrastructure, high upfront purchase price, vehicle performance, range, durability, evolving EV technology, unreliability of suppliers, availability on maintenance-service and repair ease, lack of reliable electricity sources, unknown re-sale value, details about total cost of ownership, lack of awareness, skepticism on safety and reliability, dealer reluctance to push EVs overcoming buying resistance, reliance on external sources for raw materials, wastage and recycling of battery, limited EV incentives and advertisement, and even lacking customer awareness."

Methodology

"Some scholars attempted barriers modeling using ISM but to the best knowledge of researchers, not many studies are available about EV enablers duly supported by modeling and its validation to investigate penetration in context to India."

"Mostly, scholars made use of this powerful approach for EV enabler/barrier modeling."

"If in SSIM (m, n) access is Q; then, in initial reachability matrix, it becomes 1 and (n, m) access becomes 0."

"If in SSIM (m, n) access is R; then, in initial reachability matrix, it becomes 0 and (n, m) access becomes 1."

"The enabler with the same reachability and the intersection set are at the top of the ISM model hierarchy, but these enablers do not help to attain any other element above their own level (Govindan and others [24])."

"The summation of row elements represents driving while sum of column elements (in this case enablers) represents dependence power in final reachability matrix."

Result Discussion and Validation

"To ensure better EV penetration, it is important to identify key enablers and to understand allied challenges like logical grouping and possible relationships, which is a hot research topic these days."

"In line with the research objective raised in the study, the authors not only attempted identification of enablers but also attempted its analysis, modeling, and logical grouping so as to facilitate EV penetration considering urban sprawl in Indian context."

"While this enabler may be an important enabler for a developed country, but for a country like India, it does not influence EV penetration."

"Results of BWM clearly indicate that enablers like high capacity batteries with low recharging time (EVE2), government policies and commitment to promote EVs (EVE4), subsidized purchase prices for EVs (EVE1), ensuring raw materials and resources availability for EV within country (EVE11), and development of sustainable electricity sources (EVE12) are critical for the success of EV penetration in India."

Managerial Implications

"Joint or combined efforts from government, corporate houses as private players, and from public in the form of wholehearted support will accelerate this penetration and will be game changer for addressing mobility needs and improvement in environmental parameters influencing quality of life."

"Incentives and subsidies for potential EV users in the form of tax savings, differential parking fees for EVs, exploring the feasibility of dedicated lanes, and initial pricing of EV are the enablers which should be immediately addressed to accelerate EV penetration and helping India to achieve the agreed carbon emission cuts."

"In terms of solutions leveraging these enablers; again, a multi-prong approach will be needed like installing a good number of solar charging stations, encouraging use of wind turbines along with in-house rooftop-based electricity generation to improve the electricity generation, promoting the home charging, and its increased use for EVs."

Conclusions

"This study concludes by highlighting the role of key enablers influencing EV penetration."

"It also guides both policy makers and researchers about how to use these identified enablers in a focused manner to combat with challenges pertaining to EV penetration."

"The major goal of this study is to identify and model the enablers of EV penetration in India, to understand the interactions, prioritization, and positioning between them for effective actions and policy-related alignments to mitigate the ill effects of urban sprawl on transportation."

"Majority of the studies discuss about barriers to EV penetration but overlook the role of enablers."

"This means other EVEs play a significant role in the implementation of EV penetration in India."

"In context to linkage enablers, one finds EVE5 and EVE6, falling in this category, where high driving and high dependence strengths of these two enablers clearly indicate influence the EV penetration."

Research Limitations and Future Scope

"This research lays emphasis on identifying and prioritizing EV enablers for successful penetration in context to India but it does not cover the scenarios of developed countries."

"Modeling approach like ISM has its own limitations like in ability to quantify the weights of enablers."

"The ISM model developed in this study helps to prioritize and ranks the enablers but it does not cover in determining the degree of influence of the enablers."

"Researchers may initiate comparative studies in context other developing nations like Malaysia, Indonesia, China, and Brazil and may report EV penetration status influencing urban sprawl based transportation challenges."

1.2.4 Vision Net Zero: A Review of Decarbonisation Strategies to Minimise Climate Risks of Developing Countries [25]

This is a machine-generated summary of:

Das, Ananya; Ghosh, Arpita: Vision Net Zero: A review of decarbonisation strategies to minimise climate risks of developing countries [25]

Published in: Environment, Development and Sustainability (2023).

Link to original: https://doi.org/10.1007/s10668-023-03318-6.

If you want to cite the papers, please refer to the original.

For technical reasons we could not place the page where the original quote is coming from.

Abstract-Summary

"To stop climate change, there is an urgent need to redefine our industrial and agricultural practices."

"Developing nations need proper framework and strategies to enable a decarbonisation pathway."

"We discuss performance of developing countries (taking the case of India and its environmental, social, economic backgrounds) in achieving 'net zero vision' as discussed in the Conference of Parties (COP) under UNFCCC."

"We have emphasised on: (a) sustainable practices countries need to follow, (b) key elements of the net-zero framework, and (c) key institutional strategies to support key elements of sustainability practices."

"This study helps in developing a proper framework that is essential to achieve the vision of COP 26 and COP 27 on "net zero" and the way forward for COP 28."

"Bibliometric mapping of countries shows that developed nations such as the USA and UK are way ahead in "decarbonisation" research as compared to developing nations like India and other South Asian countries."

Introduction

"According to Climate Finance and the Nationally Determined Contribution of India, India is the world's 11th largest economy in emerging countries, accounting for 33% of Certified Climate Bonds."

"It also provides credible net zero plans which include: (a)setting short-term and long-term goals, (b) being thorough in accounting for all greenhouse gas emissions, (c) continuing commitment to lowering greenhouse gas emissions, (d) planning to permanently remove carbon dioxide from the atmosphere in order to set off remaining emissions, (e) publishing a detailed plan with immediate activities to reach net zero by the middle of the century, (f) increasing climate finance, creating new financial instruments and markets, (g) building awareness and transparency about climate risks and opportunities, (h) putting in place compensatory mechanisms to address socioeconomic effects including social assistance programs, reskilling and redeployment initiatives for impacted workers and economic diversification initiatives."

Methodology

"Scopus, the world's largest database of abstracts and citations, was used to search for relevant documents using a variety of keywords in order to grasp the current state of global research and by extension, the concern of scientists related to decarbonisation and net zero."

"The Scopus database was simulated with different sets of keywords to download files for bibliometric analysis."

"We searched the database with different combination of keywords such as 'COP', 'NetZero', 'Decarbonisation', 'Emission' on 30 March, 2023, encompassing research data till December, 2022."

"The search found 17 articles with all fields criteria."

"A separate search with a single keyword like 'NetZero' or 'Decarbonisation' found 261 and 19,237 documents, respectively, with all fields criteria."

"The final search was made with the filtering criteria 'Article title' with keyword 'Decarbonisation' on Scopus database in order to keep maximum amount of relevant articles."

Recent Literature

"The research question for article search was 'What are the decarbonisation strategies to achieve net zero?'."

"According to Fankhauser and others [26], there are five attributes to achieve NetZero: (a) front-loaded emission reductions, (b) a comprehensive approach to reduce emission, (c) cautious use of carbon dioxide removal, (d) effective regulation of carbon offsets, (e) equitable transition to net zero."

"The articles showed that change to a net-zero carbon economy will be largely driven by industry and policy, which will need expertise in strategies of both emissions and renewable energy."

"It proves that researchers are recently focusing mainly on strategies to achieve Net Zero target in order to achieve the goals of Paris Conference."

"The influential article published by Bumpus and Liverman [27] from Oxford University, UK was found to be the most cited by researchers on decarbonisation and carbon offsets."

Elements and Sub-elements of Achieving Net-Zero

"Some activities in the present scenario could include: (i) Conduct studies on geological storage capacity in India. (ii) Draft and implement policies and carbon financing mechanisms. (iii) Identify suitable technology options and ascertain their potential. (iv) Provide policy/regulatory options at the national and state-levels to enable industries make an informed decision on implementing CCUS technologies. (v) Frame policies for finance/funding options (since the upfront CAPEX cost of implementation of CCUS technology is very high). (vi) Create road-map for development of sector-specific technologies and define the role of the taskforce in laying this roadmap. (vii) Create CO_2 grid to enable CO_2 supply to players in hub areas. (viii) Draft charter/document on the financial incentives to be achieved, for setting up commercial-scale CCUS initiatives. (ix) Ensure availability and quality of adequate data. (x) Facilitate utilisation and conversion of the CO_2 to ethanol/methanol/adipic acid on a larger scale. (xi) Conduct more research studies on the feasibility/costs of implementing, and the priorities. (xii) Demarcate important routes of the CCUS technology for achieving net zero in the commercial sector."

Stakeholders: Policy Actions

"Besides, it should also include management courses to build skills for the energy and power management sectors."

"For enterprises working with green labelling, green products, to get incorporated into a sustainable system, there must be proper demand to need ratio mapping in place (demand to need mapping would include mapping the social factors, geographic influences, economic factors that directly and indirectly influence communities' need for a shift to green products, and the status of green products manufacturing for that place)."

"Investment from banking, financial and insurance sectors (BFSI) is required for the construction of infrastructure, to provide loans to support sustainable growth, and to mobilise funds for green adaptation."

"Such factors in the ecosystem would include building policies, development of management and leadership infrastructure, and enough interest and engagement from producers, users and financial sectors, including banking and insurance (Bhasin & Gulati [28])."

Conclusion and Policy Implications

"India is on the route to restructuring its economy with new green developments (its paradigm), optimal land use, expanding e-mobility sectors and services, and with different techniques of afforestation, etc. India as a country can therefore provide and come up with a template norm for all other developing countries and emerging economies to follow if all the plans and structures are in place for a sustainable future with low carbon pathways to achieve net zero target (DeWit [29]; NITI Aayog report [30])."

"India's climate action plan will be focussed on the following seven areas: (a) Low-carbon electrical systems that are development-compatible; (b) Comprehensive, effective and low-carbon transportation systems; (c) Material and energy efficiency in buildings; (d) Disentangling emissions from growth and creating an effective, low-emission industrial system; (e) Engineering solutions involving CO_2 removal; (f) Increasing forest cover while taking socioeconomic and ecological factors into account; (g) Increasing climate resilience in the fight against poverty and creation of jobs."

1.2.5 Reliability Investigation of Electric Vehicle [31]

This is a machine-generated summary of:

Ammaiyappan, Bharathi Sankar; Ramalingam, Seyezhai: Reliability investigation of electric vehicle [31].

Published in: Life Cycle Reliability and Safety Engineering (2019).

Link to original: https://doi.org/10.1007/s41872-018-00071-y.

If you want to cite the papers, please refer to the original.

For technical reasons we could not place the page where the original quote is coming from.

Abstract-Summary

"Investigation of reliability is of vital importance to design, implement and manage the operation of vehicles."

"The main components of the electric vehicle (EV) are the batteries and the electric drive."

"This paper focuses on the reliability aspects of the lead acid battery, motor controller and electric motor employed for the operation of electric vehicles."

"This paper presents the simulation model using MATLAB/SIMULINK for analyzing the reliability of electric vehicles."

"This model is discussed in detail and the result of reliability estimation based on an electric vehicle is presented and analyzed."

"The results are verified experimentally with the working model of electric vehicle available in the laboratory."

Introduction

"Research activities on the reliability of components, power electronic converters and the whole drive for EVs have been reported in literature (Berthon and others [32]; Renken and others [33]; Hirschmann and others [34])."

"A test bench implemented with various driving cycles to verify the reliability of new prototypes of motor controller for electric motors in EVs is presented."

"This reliability model focuses on the effects of temperature and thermal cycle on the failure rates of the key power components of motor controllers."

"The developed electric vehicle using BLDC drive is compared with the DC motor drive-based vehicle in terms of reliability."

"To overcome the limitations of the existing methods, this paper presents a reliability model for electric vehicles for BLDC drive."

Battery Thermal Management System

"For an electric vehicle, the main heat dissipation components are the battery and the motor."

"The heat should be controlled at a certain level to get the best performance; so a thermal management system is an essential factor for the EV vehicle and improves its reliability and lifetime, and the safety of the batteries and the motor (Marchesoni and Savio [35]; Negarestani and others [36]; Hsu and others [37]; Leonhard [38])."

"A simple, fast, and effective equivalent circuit model structure for lead acid batteries was implemented to facilitate the battery model part of the system model."

"The battery model is capable of providing accurate simulation results and very fast simulation speed."

"A simple thermal model is used to model the battery temperature."

"The depth of charge is the percentage of battery capacity to which a battery is discharged."

Fault-Tolerant Issues

"Control schemes are proposed to keep the neutral point voltage balanced and get a stable and continuous output under single device fault in open-circuit and short-circuit conditions."

"The simulation results show that the inverter motor controller has fault tolerant-ability under open-circuit and short-circuit conditions."

"Without the proposed control, the converter is not able to work properly after the device failure occurs, including the unbalance of the neutral point voltage of the dc-link and unsymmetrical output voltage of the converter."

"When the fault phase has to be connected to the neutral point of the dc-link, only "0" output voltage level is available, and consequently the output voltage is reduced in the three-level conventional inverter motor controller."

Hardware Implementation of Battery Selection

"The battery bank capacity in ampere hour (AH) is given bywhere DOD is the battery depth of discharge; D_s the battery autonomy or storage days; η the temperature correction factor = 0.9; E_c(Ah) the energy or load given by ampere hour."

"Batteries in parallel are given bywhere B_{sc} is the capacity of the selection battery (Ah)."

"Batteries in series are given bywhere V_n is the nominal voltage of battery; V_s the nominal voltage of system."

Failure Rate Model

"The motor controller IGBT failure rate can be determined bywhere the first term λ_{die}, which is mainly determined by the junction temperature in the mission profile, represents the failure rate component related to the die of motor controller IGBTs; and the second term $\lambda_{package}$ denotes motor controller IGBT package failures that are caused by the number and magnitudes of thermal cycles that the devices undergoes."

"i represents the amplitude of the thermal variation that the device undergoes in the ith phase of its mission profile.(πn)i is the influence factor that is related to the annual number of the thermal cycle experienced by the package with the amplitude of i. The failure rate model further demonstrates that the junction temperature and the temperature cycle have a significant influence on the failure of the motor controller IGBTs."

Driving Cycle

"The torque–speed characteristics of vehicles versus time determine the operating conditions of the power electronic converters in the drive system, which finally affect the electrical and thermal stresses of the key power components."

"Since these driving cycles have been accepted by the industry and widely used to assess the performance of the vehicles, herein they are employed to emulate the operating patterns of HEVs."

"The differential rear axle of the vehicle is connected to the driving shaft of the BLDC motor through the gear coupling."

"The BLDC drive reduces the weight of the power transmission system resulting in lower maintenance and smooth operation of the vehicle."

"A 3 kW solar-powered electric vehicle using BLDC drive was developed."

Conclusion

"The reliable model developed in this paper included not only the thermal analysis of the battery and power converter, but also accounted for the effects of losses related to driving behaviors."

"From the reliability model, it is observed that the proposed EV using BLDC drive is more reliable and the failure rates are less compared to the conventional EV."

"On the basis of the accurate reliability analysis, improvement in design of the electric vehicle, control methods and energy management strategies can be realized to further enhance the performances of vehicles and to reduce the operation and maintenance cost."

1.2.6 Scheduling of Electric Vehicle for Performance Improvement of Electricity Distribution Network: A Case Study in Indian Scenario [39]

This is a machine-generated summary of:

Kumar, Bablu; Samantaray, Soumyakanta; Kayal, Partha: Scheduling of Electric Vehicle for Performance Improvement of Electricity Distribution Network: A Case Study in Indian Scenario [39].

Published in: Journal of The Institution of Engineers (India): Series B (2022).

Link to original: https://doi.org/10.1007/s40031-022-00824-5.

If you want to cite the papers, please refer to the original.

For technical reasons we could not place the page where the original quote is coming from.

Abstract-Summary

"Further, EV batteries can be used purposefully for ancillary services in residential electricity distribution networks."

"This paper proposes a charging and discharging scheduling approach for EVs at residential buses to improve the performance of the electrical distribution infrastructure."

"The performance of the electricity distribution system has been measured in terms of daily average real power losses of the whole network and the average voltage deviation of the buses."

"The adequacy of the proposed method is tested on a typical 28-bus Indian distribution network."

Introduction

"In [40], a day head scheduling of microgrid along with a scenario-based stochastic optimization problem characterized by MCS has been defined to minimize the total operating cost considering the charging and discharging pattern of the EVs."

"In [41], the authors have presented a strategy to optimize the charging and discharging scheduling of the EVs, considering the objective of network power loss minimization."

"Papers, either power loss minimization or voltage profile improvement of the buses in the distribution network was considered as the objective for charging and discharging scheduling of the EVs."

"Minimization of network power loss and deviation in voltage magnitude of the buses in the distribution system are considered simultaneously for charging and discharging scheduling of the EVs."

"A complex problem has been developed to prosper the power distribution efficiency by strategic charging and discharging of the EVs."

Problem Formulation

"This study aims to maximize the performance of the distribution network from the viewpoint of minimization of average network power losses and minimization of average voltage deviation."

"This happens due to variation in power demand throughout the day at the different load buses in the network."

"Power loss of the network varies throughout the day."

"Power demand at the buses varies due to charging/discharging of the EVs, which further causes to vary power losses in the network."

"The network's daily average real power loss can be realized as Here, N is the total number of buses in the network, and nT is the total number of time segments in a day."

"During charging hours, battery of the EV takes power from the network in a lightly loaded condition."

"During discharging hours, EVs feedback power into the network when the demand is high."

Methodology

"In each flight, particles change their position and velocity and attempt to meet in an additional encouraging area of the solution."

"Each particle "" is related to two vectors, i.e., velocity vector and position vector."

"Through iteration, the development of particles advances to an optimal or close to an optimal solution."

"PSO and GOA optimization techniques are applied to the scheduling problem to obtain the optimum value of the objective function."

Result and Discussion

"It is considered that different numbers of EVs will be connected at five different residential load points in the distribution network and follow similar charging and discharging pattern."

"The remaining time on the weekend days can be utilized for charging/discharging the EVs."

"Suitable charging and discharging pattern are identified through PSO- and GOA-based simulation algorithms for the purposeful use of the EVs."

"PSO-based approach was able to generate the solution in 44,345 s. The EVs are used for transportation in the time slot of 10 to 17 and idle for charging and discharging."

"It is seen from the results that the proposed approach is able to minimize the voltage deviation in a better way, along with a similar reduction in real power losses."

"It is to be noted that the EVs are idle for charging and discharging from 17 to 21 h slots in the weekend scenario."

Conclusion

"Suitable charging and discharging slots for the EVs have been identified utilizing PSO- and GOA-based approaches."

"It has been observed that the interaction of EVs with the grid is different on weekdays and weekend days."

"The optimized charging and discharging time slots for the weekdays and weekend days are different."

"The study considers a number of constraints and available time slots for charging/discharging in the Indian context, which may differ with the change in region or country."

"The proposed approach is compatible with other constraints and degrees of preference toward different objectives."

"The study can be further extended with the provision for distributed scheduling in which charging/discharging time slots may be different at different buses in the network."

Bibliography

1. Sun, Dongying; Kyere, Francis; Sampene, Agyemang Kwasi; Asante, Dennis; Kumah, Naana Yaa Gyamea An investigation on the role of electric vehicles in alleviating environmental pollution: evidence from five leading economies. Environ Sci Pollut Res (2022). https://doi.org/10.1007/s11356-022-23386-x
2. Busu M, Nedelcu AC (2021) Analyzing the renewable energy and CO_2 emission levels nexus at an EU level: a panel data regression approach. Processes 9(1):130. https://doi.org/10.3390/PR9010130
3. Sampene AK, Li C, Oteng-Agyeman F, Brenya R (2022) Dissipating environmental pollution in the BRICS economies: do urbanization, globalization, energy innovation, and financial development matter? Environ Sci Pollut Res 1:1–21. https://doi.org/10.1007/S11356-022-21508-Z/TABLES/3
4. Westerlund J (2007) Testing for error correction in panel data. Oxford Bull Econ Stat 69(6):709–748. https://doi.org/10.1111/J.1468-0084.2007.00477.X
5. Chudik A, Pesaran MH (2015) Common correlated effects estimation of heterogeneous dynamic panel data models with weakly exogenous regressors. J Econom 188(2):393–420. https://doi.org/10.1016/J.JECONOM.2015.03.007
6. Baydoun H, Aga M (2021) The effect of energy consumption and economic growth on environmental sustainability in the GCC countries: does financial development matter? Energies 14(18):5897. https://doi.org/10.3390/EN14185897

7. Huang SZ, Chien F, Sadiq M (2021) A gateway towards a sustainable environment in emerging countries: the nexus between green energy and human Capital. Econ Res-Ekonomska Istrazivanja. https://doi.org/10.1080/1331677X.2021.2012218
8. Teal F, Eberhardt M (2010) Productivity analysis in global manufacturing production. Economics Series Working Papers. https://ideas.repec.org/p/oxf/wpaper/515.html . Accessed 15 June 2022
9. Pesaran MH (2007) A simple panel unit root test in the presence of cross-section dependence. J Appl Economet 22(2):265–312
10. Dumitrescu EI, Hurlin C (2012) Testing for Granger non-causality in heterogeneous panels. Econ Model 29(4):1450–1460. https://doi.org/10.1016/J.ECONMOD.2012.02.014
11. Ahmad M, Ahmed Z, Yang X, Hussain N, Sinha A (2022) Financial development and environmental degradation: do human capital and institutional quality make a difference? Gondwana Res 105:299–310. https://doi.org/10.1016/J.GR.2021.09.012
12. Gupta A, Garg A (2022) Modelling the enablers for adoption of electric vehicles in India. Int J Syst Assur Eng Manage (2022). https://doi.org/10.1007/s13198-022-01687-1
13. IEA (2019) Scaling-up the transition to electric mobility. Global EV Outlook 2019, IEA, Paris. Retrieved from https://www.iea.org/reports/global-ev-outlook-2019
14. Eberle U, Von Helmolt R (2010) Sustainable transportation based on electric vehicle concepts: a brief overview. Energy Environ Sci 3(6):689–699
15. WRI India (2019) A guidance document on accelerating electric mobility in India [Ebook], pp 1–104. Retrieved from https://wriindia.org/sites/default/files/Accelerating%20electric%20mobility%20in%20India_WRI%20India_CBEEVIITM.pdf
16. Krause RM, Carley SR, Lane BW, Graham JD (2013) Perception and reality: public knowledge of plug-in electric vehicles in 21 US cities. Energy Policy 63:433–440
17. Namdeo A, Tiwary A, Dziurla R (2014) Spatial planning of public charging points using multi-dimensional analysis of early adopters of electric vehicles for a city region. Technol Forecast Soc Chang 89:188–200
18. Joshi AM, Ezzat H, Bucknor N, Verbrugge M (2011) Optimizing battery sizing and vehicle lightweighting for an extended range electric vehicle (No. 2011-01-1078). In: SAE 2011 World Congress & Exhibition, Detroit Michigan, United States. https://doi.org/10.4271/2011-01-1078
19. Liao F, Molin E, Wee BV (2017) Consumer preferences for electric vehicles: a literature review. Transp Rev 37(3):252–275
20. Rehman MA, Seth D (2023) Investigation and modeling of electric vehicle enablers (EVE) for successful penetration in context to India: mitigating the effect of urban sprawl on transportation. Environ Sci Pollut Res (2023). https://doi.org/10.1007/s11356-023-26022-4
21. Tarei PK, Chand P, Gupta H (2021) Barriers to the adoption of electric vehicles: evidence from India. J Clean Prod 291. https://doi.org/10.1016/j.jclepro.2021.125847
22. Patyal SV, Kumar R, Singh Kushwah S (2021) Modeling barriers to the adoption of electric vehicles: an Indian perspective. Energy 237. https://doi.org/10.1016/j.energy.2021.121554
23. Goel S, Sharma R, Rathore AK (2021) A review on barrier and challenges of electric vehicle in India and vehicle to grid optimisation. Transp Eng 4(February). https://doi.org/10.1016/j.treng.2021.100057
24. Govindan K, Muduli K, Devika K, Barve A (2016) Investigation of the influential strength of factors on adoption of green supply chain management practices: an Indian mining scenario. Resour Conserv Recycl 107:185–194. https://doi.org/10.1016/j.resconrec.2015.05.022
25. Das A, Ghosh A (2023) Vision net zero: a review of decarbonisation strategies to minimise climate risks of developing countries. Environ Dev Sustain (2023). https://doi.org/10.1007/s10668-023-03318-6
26. Fankhauser S, Smith SM, Allen M, Axelsson K, Hale T, Hepburn C, Kendall JM, Khosla R, Lezaun J, Mitchell-Larson E, Obersteiner M, Rajamani L, Rickaby R, Seddon N, Wetzer T (2022) The meaning of net zero and how to get it right. Nat Clim Chang 12(1):15–21
27. Bumpus AG, Liverman DM (2008) Accumulation by decarbonization and the governance of carbon offsets. Econ Geogr 84(2):127–155. https://doi.org/10.1111/j.1944-8287.2008.tb00401.x

28. Bhasin NK, Gulati K (2021) A study of the readiness of Indian banks to absorb COVID-19s impact through new emerging technologies and strategies for competitive advantage. In: E-collaboration technologies and strategies for competitive advantage amid challenging times, pp 50–75
29. DeWit A (2022) Climate code red" at COP26 and the diversification of decarbonization narratives
30. HEALTHY STATES PROGRESSIVE INDIA Report on the Ranks of States and Union Territories, NITI Aayog report. (2020). https://www.niti.gov.in/sites/default/files/2021-12/NITI-WB_Health_Index_Report_24-12-21.pdf
31. Ammaiyappan BS, Ramalingam S (2019) Reliability investigation of electric vehicle. Life Cycle Reliab Saf Eng (2019). https://doi.org/10.1007/s41872-018-00071-y
32. Berthon A, Gustin F, Bendjedia M, Morelle J, Coquery G (2009) Inverter components reliability tests for hybrid electrical vehicles. In: Record of IEEE power electronics and motion control conference, pp 763–768
33. Renken F, Ehbauer G, Karrer V, Knorr R, Ramminger S, Seliger N, Wolfgang E (2007) Reliability of high temperature inverters for HEV. In: Record of power conversion conference, pp 563–568
34. Hirschmann D, Tissen D, Schroder S, De Doncker R (2007) Reliability prediction for inverters in hybrid electrical vehicles. IEEE Trans Power Electron 22:2511–2517
35. Marchesoni M, Savio S (2005) Reliability analysis of a fuel cell electric city car. In: Record of European conference on power electronics and applications, pp 1–10
36. Negarestani S, Ghahnavieh A, Mobarakeh A (2012) A study of the reliability of various types of the electric vehicles. In: Record of IEEE international electric vehicle conference, pp 1–6
37. Hsu JS, Nelson SC, Jallouk PA et al (2005) Report on Toyota Prius motor thermal management. Oak Ridge National Laboratory, Technical Report, ORNL/TM-2005/33
38. Leonhard W (2001) Control of electrical drives, 3rd edn. Springer, Berlin
39. Kumar B, Samantaray S Kayal P (2022) Scheduling of electric vehicle for performance improvement of electricity distribution network: a case study in Indian scenario. J Inst Eng (India): Ser B. https://doi.org/10.1007/s40031-022-00824-5
40. Li H, Rezvani A, Hu J, Ohshima K (2021) Optimal day-ahead scheduling of microgrid with hybrid electric vehicles using MSFLA algorithm considering control strategies. Sustain Cities Soc 66:102681. https://doi.org/10.1016/j.scs.2020.102681
41. Usman M, Tareen WUK, Amin A, Ali H, Bari I, Sajid M, Seyedmahamoudian M, Stojcevski A, Mahmood A, Mekhilef S (2021) A coordinated charging scheduling of electric vehicles considering optimal charging time for network power loss minimization. Energies 14(5336):1–16
42. Lieven T (2015) Policy measures to promote electric mobility—a global perspective. Transp Res Part A Policy Pract 82:78–93

Part II
Upstream

Part II contains chapters 2 and 3. Since the battery is the most important component in the electric vehicle (EV), chapter 2 discusses the upstream value chain of the battery chain, and focuses on the minerals needed. Chapter 3 discusses the technology used to produce EV batteries.

Chapter 2
Sourcing Minerals for the Battery

2.1 Introduction by Editor

The objective of the chapter is to understand what kind of critical role minerals play in the EV value chain. This chapter will discuss the various aspects of lithium, Nickel Cobalt, Copper, Graphite and other critical minerals sources for EV batteries in India. The battery of the EV uses various minerals like Lithium, Nickel and Cobalt. Lithium is a key component of batteries and other electronics. Cobalt is difficult to substitute—it has some properties that make EV batteries stable. In EVs, the battery accounts for 70% of the cost of two-wheelers and 50% of cars. All EV production depends on the supply of cobalt and lithium from a few countries. 65% of lithium reserves are in Bolivia and Chile, while 60% of cobalt reserves are in Congo. For all countries, securing a supply of essential metals is critical. From India's point of view, the lack of critical minerals has been pointed out. Some lithium is reported to have been found in Jammu and Kashmir, though it may take some time to be commercially available.

2.1.1 Introduction

The most important component of the EV is the battery. There are several kinds of batteries that are in use. However, the most prominent is the lithium-ion (or Li-ion). The battery of the EV is made from several critical minerals like Lithium, Cobalt, Nickel, and Manganese using complex processes. This chapter will discuss the various aspects of lithium and cobalt and other critical minerals sources for EV batteries in India.

The value chain of the EV will be discussed in a subsequent chapter. The alternative battery technologies will be discussed in Chap. 3.

S. Ray and D. Mukherji (eds.), *A Primer on Electric Vehicles in India*,
https://doi.org/10.1007/978-981-97-8966-5_2

2.1.2 Critical Minerals [1] in the Battery

The minerals that are used in an EV battery include lithium, nickel, cobalt, manganese etc. For all countries, securing a supply of essential metals is critical. All EV production depends on the supply of cobalt and lithium from a few countries (IEA, 2022). Lithium is a key component of batteries and other electronics. The major use of cobalt is in chemicals (paints and dyes) and corrosion-resistant alloys.

In EVs, the battery accounts for 70% of the cost of two-wheelers and 50% of cars. The cost of EV batteries (though they have decreased) is mainly due to expensive metals like cobalt, nickel etc.

2.1.2.1 Lithium

Lithium is a Soft and silvery-white alkali metal. It is the lightest metal with atomic number 3 and the lightest solid element. It is soft—and can be cut with a knife. It has very low density. It is highly reactive and flammable in contact with air, therefore must be stored in mineral oil. It is never found free in nature but exists in combined forms in ocean water or in brines (and obtained by electrolysis of lithium chloride and potassium chloride). Lithium can float on even the lightest hydrocarbon oils and is one of only three metals to float on water (the other two are sodium and potassium).

The use of electrodialysis and electrochemical intercalation has been proposed to extract lithium from seawater. These sources are less expensive to mine than from rock such as spodumene, petalite, and other lithium-bearing minerals.

India has almost no Lithium mining at present though there have been recent attempts to map various regions for the metal. Researchers at the Atomic Minerals Directorate (under India's Atomic Energy Commission) have estimated lithium reserves of 14,100 tonnes in a small patch of land surveyed in Southern Karnataka's Mandya district recently. The exploratory work is meanwhile also on to possibly extract lithium from the brine pools of Rajasthan and Gujarat and the mica belts of Odisha and Chhattisgarh.

The prices of lithium have fallen dramatically in the last few years to justify the use of lithium in EV batteries. Lithium is priced via quarterly contracts that are sometimes announced to the market.

According to geologists, Rare Earths Elements (REEs) are not rare, but they are precious. Comprising 17 elements that are used extensively in both consumer electronics and national defence equipment, rare earth elements (REEs) were first discovered and put into use in the United States. However, production gradually shifted to China, where lower labour costs, less concern for environmental impacts, and generous state subsidies enabled China to corner most of the global production in such materials.

2.1.2.2 Cobalt

Cobalt is a grayish-white metal and is hard and brittle. It makes a strong alloy with other metals, with a high melting point and low conductivity. Cobalt is predominantly used for lithium-ion batteries and for producing brilliant blue pigments [2] Cobalt is difficult to substitute—it has some properties that make EV batteries stable. Cobalt is primarily produced as a by-product of copper and nickel mining, though a smaller portion is extracted from primary cobalt mines. Cobalt is also sourced from the recycling of cobalt-containing materials, such as batteries, aerospace alloys, and industrial waste. Recycling plays an important role in cobalt supply, as it helps to reduce the reliance on primary production and supports circular economy principles [3]. The majority of global cobalt mines are in the Democratic Republic of Congo (DRC), while other significant cobalt-producing countries include Russia, Australia, Canada, and the Philippines. Cobalt pricing is less transparent and is not announced publicly and long term deals are based on volume.

2.1.2.3 Nickel

Nickel is a silvery-white metal found naturally in the earth's crust and it has chemical properties that enable it to form multiple chemical compounds. Since pure nickel is hard yet ductile and pliable, it is used as a strengthening component in metal alloys. It is also an excellent conductor of both heat and electricity [4]. Nickel is widely used in manufacturing stainless steel for its corrosion resistance properties. The usage of Nickel in batteries increases the battery charge density and improves storage capacities. Major suppliers of Nickel are in Russia, Indonesia, Australia, South Africa, and Canada [5, 6].

[1] What Does Cobalt Look Like?—Great American Adventures (great-american-adventures.com

[2] https://geologyscience.com/ore-minerals/cobalt-ore/

[3] https://sites.dartmouth.edu/toxmetal/more-metals/nickel-hidden-in-plain-sight/the-facts-on-nickel/

2.1.2.4 Copper

Copper is the most critical raw material for the electric industry. Copper will play a critical role in the drive to 'net zero': the $150 trillion global shift to electrify and decarbonize the world by 2050 in an effort to slow climate change. The global copper industry is about to face a supply crisis, due to a lack of mining investment and political tensions in key copper regions. Since 2018, copper reserves have dropped by 80% and continue to decline. By 2030, global shortage of around 19 million tonnes of copper is expected. Copper is the only viable option for conducting electricity efficiently, with no currently known alternatives. S&P Global projects that roughly 50 million tonnes of the metal by 2035 will be needed. (Bloomberg, 2022).

EVs alone consume as much as 80 kg of copper per vehicle, or four times more than a conventional car. With net zero requiring at least 350 million EVs on the road by 2030, the global demand would be 28 million tons from the EV industry alone. US and China are planning to spend a combined US$550 billion on transmission infrastructure which will need a huge amount of copper. The major Copper producers as of today are: (in Thousand Tons) 1. Chile 5700, 2. Peru 2200, 3. China 1700 and 4. Congo, Democratic Republic 1300.

2.1.2.5 Graphite

Graphene is a naturally occurring substance in graphite which is considered one of the strongest substances found on the surface of the earth. It is 200 times stronger than steel. Graphene itself is a very good conductor of heat and electricity. Graphite is the naturally occurring form of carbon. It is a stable form of it. The usage of graphite is wide and varied. The usage of graphite is extensive—from pencils to nuclear plants.—the usage is highly extensive. It has its usage as electrodes, in brake linings, in batteries, and in the process of manufacturing steel.

The Anode use of Graphite in Li-Ion batteries makes it one of the most needed minerals for EV needs. India is one of the leading graphite producers in the world. According to the data of 2013, India has reserves of about 11m tonnes. India's graphite occurrences are found in states like Jammu and Kashmir, Gujarat, Jharkhand, Arunachal Pradesh, Karnataka, Kerala, Maharashtra, Tamil Nadu, Odisha, Chhattisgarh and Rajasthan. However, the deposits which can be of economic significance are found in Andhra Pradesh, Chhattisgarh and Arunachal Pradesh. Some of these fields are yet to be exploited. As far as graphite mining and processing is concerned, Jharkhand, Odisha and Tamil Nadu are the major states where mining is being undertaken. Lately, Graphite mining has begun in Arunachal Pradesh as well (Fig. 2.1).

The graphite production in India 2021 65,000 MT.

2.1.3 Imports of Raw Materials

India is dependent on imports of both cobalt (particularly cobalt oxide) and lithium (oxide and hydroxide) [7]. Using data from DGFT shown in Table 2.1, we note the following:

The imports of nickel are significant. Other minerals imported include graphite and lithium. A significant amount of lithium-ion cells are also imported, which we discuss in Chaps. 3 and 5

repeated below. See https://pib.gov.in/PressReleasePage.aspx?PRID=1936083. See also https://csep.org/working-paper/critical-minerals-for-india-2023/

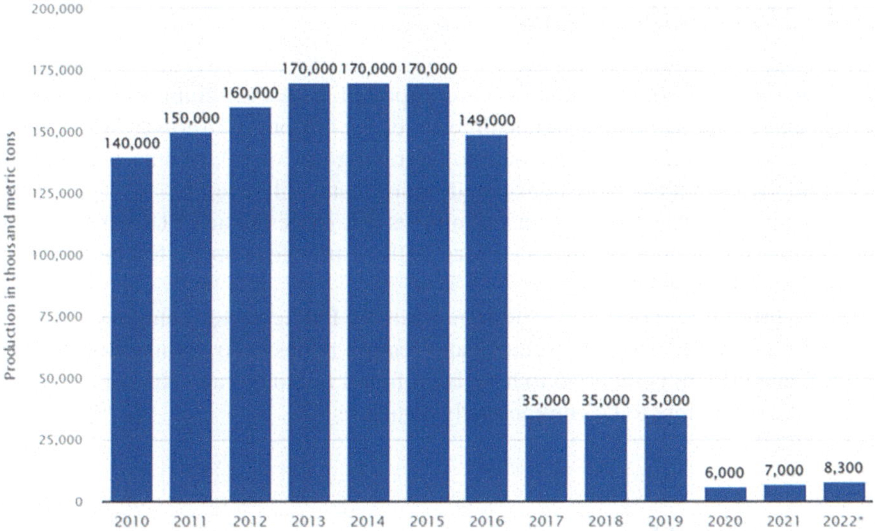

Fig. 2.1 Graphite production in India (*Source* Graphite production India Statista, 2023)

Table 2.1 Raw materials that are imported by India (Value in Rs. lakhs)

HS	Description	2018–19	2019–20	2020–21	2021–22
750,210	Nickel, not alloyed	361,350.8	336,431.4	313,178.9	548,389
380,110	Artificial graphite	21,284.45	18,033.86	18,398.43	32,398.41
282,739	Other chloride; NES	29,329.84	27,162.46	29,413.53	36,366.54
282,520	Lithium oxide and hydroxide	21,963.53	13,085.10	12,336.18	21,612.84
280,519	Other alkali/alkaline—earth metals	9,540.74	7,661.57	6,364.21	12,771.39
750,400	Nickel powders and flakes	5,422.88	5,622.20	5,767.21	10,073.40
282,200	Cobalt oxide and hydroxde, comrcl coblt oxide	5,270.36	5,411.18	4,133.10	7,487.55
283,691	Lithium carbonates	10,845.89	7,945.26	6,341.14	6,482.45
282,690	Other complex fluorine salts	4,632.17	3,949.05	2,468.57	5,652.43
850,750	Nickel-metal hydride	4,041.29	2,953.07	3,444.78	5,279.09
750,220	Nickel alloys	4,042.88	3,578.12	8,964.37	4,846.34

Source Ministry of Commerce and Industries India

2.1.4 Minerals that Are Imported

Raw materials like lithium and cobalt are imported. 65% of lithium reserves are in Bolivia and Chile, while 60% of cobalt reserves are in Congo. However, while there are concerns regarding cobalt (related to its mining in Congo), lithium is available more widely. Both these were considered minor metals till recently.

India has been proactive in securing supplies [8]. Three Indian PSUs have set up a JV to mine for lithium and cobalt overseas. A three-pronged approach is being taken especially for lithium [9].

India currently imports all its lithium needs. A Public sector company "Khanij Bidesh India" has been set up to source and acquire mines in Argentina, Bolivia and Chile. Some lithium is reported to have been found in Jammu and Kashmir, though it may take some time to be commercially available [10].

2.1.5 Challenges

Unfortunately, India has a factor disadvantage since there are no mines of Lithium, Cobalt, Nickel, or Manganese in the country. These materials are essential for making the cathode of a cell. On the anode side graphite is available but it needs to be processed to make cell grade. The cell-grade graphite can be manufactured from by-products of steel.

Even though India is expected to secure material supplies [11] from countries like Chile, Argentina, Australia etc., it is dependent on China for refined battery-grade materials [12]. This leads to the question of India's critical mineral security with China [13]. Always a long-term planner, China has secured a supply of essential metals with purchases of mines in Congo, Bolivia, Chile, and Australia. It controls half the cobalt mines in Congo. India has been proactive in securing lithium. However, cobalt is more difficult to secure. Alternatives can help. More research partnerships and investment are needed to fulfill these objectives [14].

One of the ways in which the problem of critical minerals can be solved is through the recycling of batteries. There are many ways in which this can be done as is discussed below.

[8] https://www.livemint.com/industry/energy/inside-india-s-race-to-secure-lithium-11634574267391.html

[9] https://www.livemint.com/companies/news/three-indian-psus-set-up-jv-to-mine-for-lithium-cobalt-overseas-1564674503318.html

https://www.wionews.com/india-news/india-makes-moves-to-tackle-chinese-dominance-in-lithium-market-355490

[10] https://economictimes.indiatimes.com/industry/renewables/how-lithium-reserves-can-speed-up-indias-ev-dream/articleshow/98226866.cms

[11] https://www.livemint.com/news/india/explained-new-lithium-deposits-to-help-india-reduce-dependency-on-china-heres-how-11683559672443.html#:~:text=After%20the%20first%20lithium%20reserve,import%201%2C200%20tonnes%20of%20lithium.

Machine Generated

Sardar and Mubashir (2012) contend that the lack of domestic resources becomes a constraint when large scale production of lithium-based batteries commences in India. They examine the alternatives for India and observe that effort is needed for fuel-cell type and electrically rechargeable Zn-air batteries. Ranadive and Jawadanand (2019) examine India's minerals and conclude that it needs to focus on the domestic exploration of strategic minerals and also to secure these resources through the strategic acquisition of overseas mines and signing diplomatic and trade agreements. Yano and Takanori (2015) examine the rare earth recovery potential for end-of-life hybrid transmissions. Vishvakarma and Dhawan (2019) discuss an environment-friendly process that includes indigenous heat treatment of active material, selective dissolution of lithium carbonate, and subsequently wet magnetic separation to recover cobalt values in a very short time. Anwani and Methekar (2020) examine two processes in cobalt oxalate from spent lithium-ion batteries and observe that the acidic leaching route is relatively greener and more economical than the acidic baking route. Cabral-Neto et al. (2022) study scrap batteries in the context of Brazil and conclude that number of scrap batteries generated is entirely out of step with vehicle sales data, proving the delayed effect on the generation of these residues (rebound effect).

References

Bhaskar, U. (2021) Inside India's race to secure lithium. Livemint. October 18. Link: https://www.livemint.com/industry/energy/inside-india-s-race-to-secure-lithium-11634574267391.html

Bloomberg (2022) Copper Demand Expected to Nearly Double by 2035. August 3. Link: https://www.bloomberg.com/press-releases/2022-08-03/copper-demand-expected-to-nearly-double-by-2035

Chadha, R., Sivamani, G., and Bansal, K. (2023) Critical Minerals for India 2023. CSEP working paper. Link: https://csep.org/working-paper/critical-minerals-for-india-2023/

Dartmouth Toxic Metals (n.d.) Facts on Nickel. Link: https://sites.dartmouth.edu/toxmetal/more-metals/nickel-hidden-in-plain-sight/the-facts-on-nickel/

Elements by Visual Capitalist Exploring the Material World (n.d.) Website: https://elements.visualcapitalist.com/

Emont, J. (2023) EV Makers Confront the 'Nickel Pickle'. Wall Street Journal. June 4. Link: https://www.wsj.com/articles/electric-vehicles-batteries-nickel-pickle-ind

onesia-9152b1f#:~:text=In%20the%20electric%2Dvehicle%20business,though%2C%20is%20particularly%20environmentally%20unfriendly

IEA (2023) Global EV Outlook 2023 Catching up with climate ambitions. Report. https://www.iea.org/reports/global-ev-outlook-2023

IEA (2022) The Role of Critical Minerals in Clean Energy Transitions. Report. Link: https://www.iea.org/reports/the-role-of-critical-minerals-in-clean-energy-transitions

Hendrix, C. (2023) The US strategy on critical minerals needs clearer priorities. Peterson Institute for International Economics. Blog. August 1. Link: https://www.piie.com/blogs/realtime-economics/us-strategy-critical-minerals-needs-clearer-priorities#:~:text=US%20resources%20should%20prioritize%20reducing,or%20single%20supplier%2Dbased%20disruptions.

Livemint (2023) Explained: New Lithium Deposits to Help India to Reduce Dependency on China. May 8. Link: https://www.livemint.com/news/india/explained-new-lithium-deposits-to-help-india-reduce-dependency-on-china-heres-how-11683559672443.html#:~:text=After%20the%20first%20lithium%20reserve,import%201%2C200%20tonnes%20of%20lithium.

Mat, M. (2023) Cobalt (Co) Ore. Link: https://geologyscience.com/ore-minerals/cobalt-ore/

Ministry of Mines (2023) Union Minister Pralhad Joshi unveils List of "Critical Minerals for India". PIB. June 28. Link: https://www.pib.gov.in/PressReleasePage.aspx?PRID=1936083

Morenhout, T., Kohli, P. C., Goel, S., Ray, S., Janardhanan, N., and Olsen, S. H. (2023) Securing Critical Minerals Supply Chains for the Clean Energy Transition. May 2023, T20 Policy Brief. Task Force 4 Refuelling Growth: Clean Energy and Green Transitions.

Link: https://www.orfonline.org/research/securing-critical-minerals-supply-chains-for-the-clean-energy-transition/

Newton, C. (2022) What is the Natural State of Cobalt? November 28. Link: https://great-american-adventures.com/what-is-the-natural-state-of-cobalt/

Rajshekhar, M. (2023) Critical Mineral Imports: The Devil Is in The Details. Carboncopy July 19, Link: https://carboncopy.info/critical-mineral-imports-the-devil-is-in-the-details/

The Economic Times (2023) How lithium reserves can speed up India's electric vehicle dream. February 25. Link: https://economictimes.indiatimes.com/industry/renewables/how-lithium-reserves-can-speed-up-indias-ev-dream/articleshow/98226866.cms

Varadhan, S. (2019) Three Indian PSUs set up JV to mine for lithium, cobalt overseas. Livemint. August 1. Link: https://www.livemint.com/companies/news/three-indian-psus-set-up-jv-to-mine-for-lithium-cobalt-overseas-1564674503318.html

WION (2021) India makes moves to tackle Chinese dominance in Lithium market, January 9. Link: https://www.wionews.com/india-news/india-makes-moves-to-tackle-chinese-dominance-in-lithium-market-355490

Footnotes/ Endnotes

[1] Critical minerals have been defined by IEA (2023) as "essential components in many of today's rapidly growing clean energy technologies."

According to geologists, Rare Earths Elements (REEs) are not rare, but they are precious. Comprising 17 elements that are used extensively in both consumer electronics and national defence equipment, rare earth elements (REEs) were first discovered and put into use in the United States. However, production gradually shifted to China, where lower labour costs, less concern for environmental impacts, and generous state subsidies enabled China to corner most of the global production in such materials.

[2] What Does Cobalt Look Like?—Great American Adventures (great-american-adventures.com.

[3] https://geologyscience.com/ore-minerals/cobalt-ore/

[4] https://sites.dartmouth.edu/toxmetal/more-metals/nickel-hidden-in-plain-sight/the-facts-on-nickel/

[5] elements.visualcapitalist.com.

[6] See Emont (2023).

[7] India has recently announced a list of 30 critical minerals. See https://pib.gov.in/PressReleasePage.aspx?PRID=1936083. See also https://csep.org/working-paper/critical-minerals-for-india-2023/

[8] https://www.livemint.com/industry/energy/inside-india-s-race-to-secure-lithium-11634574267391.html

[9] https://www.livemint.com/companies/news/three-indian-psus-set-up-jv-to-mine-for-lithium-cobalt-overseas-1564674503318.html
https://www.wionews.com/india-news/india-makes-moves-to-tackle-chinese-dominance-in-lithium-market-355490

[10] https://economictimes.indiatimes.com/industry/renewables/how-lithium-reserves-can-speed-up-indias-ev-dream/articleshow/98226866.cms

[11] See Rajshekhar (2023).

[12] https://www.livemint.com/news/india/explained-new-lithium-deposits-to-help-india-reduce-dependency-on-china-heres-how-11683559672443.html#:~:text=

After%20the%20first%20lithium%20reserve,import%201%2C200%20tonnes%20of%20lithium.

[13] Many countries are grappling with this issue: See e.g. Hendrix (2023).

[14] See Morenhout et al. (2023).

2.2 Machine Generated Summaries

Disclaimer: The summaries in this chapter were generated from Springer Nature publications using extractive AI auto-summarization: An extraction-based summarizer aims to identify the most important sentences of a text using an algorithm and uses those original sentences to create the auto-summary (unlike generative AI). As the constituted sentences are machine selected, they may not fully reflect the body of the work, so we strongly advise that the original content is read and cited. The auto generated summaries were curated by the editor to meet Springer Nature publication standards. To cite this content, please refer to the original papers.

Machine generated keywords: lib, metal, lithium, lithiumion, lithiumion battery, cobalt, nickel, ton, contain, recovery, material, acid, hevs, strategic, route.

2.2.1 Strategic Minerals in India: Present Status and Future Challenges [1]

This is a machine-generated summary of:

Randive, Kirtikumar; Jawadand, Sanjeevani: Strategic minerals in India: present status and future challenges [1].

Published in: Mineral Economics (2019).

Link to original: https://doi.org/10.1007/s13563-019-00189-0.

If you want to cite the papers, please refer to the original.

For technical reasons we could not place the page where the original quote is coming from.

Abstract-Summary

"The paucity of strategic minerals in India has been a serious concern for the last several decades."

"Growing global demand, high value, declining production and constant pressure for finding new mineral deposits captivate not only India's but also the world's outlook towards strategically important minerals."

"India is highly dependent on imports for the supply of strategic minerals."

"It is a major area of concern for India, since regular supply from foreign countries may be interrupted for obvious reasons such as changing the geopolitical situation or the foreign mineral producers forming cartels to dictate the prices."

"The supply of these minerals should be secured through the strategic acquisition of overseas mines by forming diplomatic ties and trade agreements."

Introduction

"That the security of a nation depends on its mineral supplies that are necessary to cater to its defense needs."

"Too much dependency on mineral supply in the modern era has evolved into a new definition of mineral trade, "A war of the minerals, by the minerals, for the minerals" underscoring their importance for national security during wartime."

"Strategic minerals are of vital importance for the economic regeneration and national security of a country."

"The concept of critical and strategic mineral resources is not new and traces its history to almost a century back to World War I when the warring nations began to experience severe shortages of materials required to sustain their war effort."

"For making new materials, a huge amount of raw minerals are required, which creates a constant demand for the mineral supplies."

Strategic Minerals Scenario

"Weston [2] maintained that a strategic mineral is a mineral that would be needed to supply the military, industrial, and essential civilian needs for the country during a national emergency."

"Kessel [3] on the other hand mentioned that the strategic minerals can only be defined by their essentiality to the national defense, their high degree of import reliance in wartime, need for their strict conservation, and distribution control."

"By any of the above definitions, strategic minerals are of great significance in national security as well as welfare of the society."

Availability, Requirement, and Utility

"The prosperity, power, and national security of a nation are by and large influenced by the availability of natural resources."

"Availability of strategic minerals at competitive prices is essential for sustaining the global economy and making substantial advancement in high-technology products, clean energy technology, and commercializing new inventions (Simandl and others [4])."

"The utility of strategic minerals is rather multifold; the greatest advantage is that technological innovations largely make use of such new materials."

"Strategic minerals are significant in defense (platforms, weapons, sensors, communications, computers, intelligence, surveillance, reconnaissance systems), aerospace (lightweight alloys), electronics industry (integrated circuits, flat screens, mobile phones) and clean energy technologies (wind turbines, electric vehicles, photovoltaic cells, fluorescent lighting, grid storage batteries, magnetic refrigeration and nuclear power etc. (Jawadand and others [5])."

"The demand and supply of strategic minerals varied widely at different periods."

"This paper presents the current strategic minerals scenario and its implications on the Indian economy."

Deficiency and Dependence of Strategic Materials

"Despite being endowed with rich mineral resources, India's import dependency for minerals is next only to oil."

"Molybdenum, tungsten, and cobalt have limited geological availability in India and have been categorized as scarce minerals (Planning Commission [6])."

"Considering the world statistics, India holds only 3% of the world reserves of rare earth as compared to 36% by China and 13% by the USA (Pandey [7])."

"India has about 70% of world's monazite reserve, it is mostly unutilized and the world is solely reliant on China's monopoly in rare earth supply."

"The strategic minerals with lower resources in India as well as scarce globally include cobalt, tungsten, lithium, and tin among others."

"Another aspect is that India will have to be a moot spectator to the games being played by resourceful nations, by controlling the supply and price of these minerals by forming cartels such as OPEC."

Policy Initiatives and Recommendations for Strategic Planning

"The concept of strategic or critical minerals is not quite new to the policy-makers in India as compared to other major economies of the world."

"A comprehensive policy ensuring adequate supplies of strategic minerals that will effectively address the issues of national security, high-tech manufacturing, and identifying areas for improvement should be framed."

"On July 4, 2016, the Ministry of Mines has announced National Mineral Exploration Policy (NMEP) which recognizes "critical minerals for industry and strategic minerals vital for national security," and recommends priority-based exploration from GSI (Gupta and others [8]; Randive and others [9])."

"This is a step in the right direction for achieving security for important mineral commodities."

Present and Future Challenges

"Strategic minerals are crucial in the production of a range of high-tech and clean energy technologies."

"The development of new products and technologies in India has boosted the demand for high-technology metals such as rare earth and cobalt."

"The worldwide demand for strategic minerals for clean energy production grows exponentially from 1971 to 2011 (Speirs and others [10])."

"Rare earth, in particular, would witness a significant increase in demand, due to India's commitment to renewable energy and a growing emphasis on green technologies (Khanna and Ganeshan [11])."

"India will face scarcity of the minerals necessary for clean-energy products and for manufacturing high-technology products."

"Supply disruption risks are high because the production of minerals is concentrated in a limited number of countries (De Ridder [12])."

"Increasing demand instigates competitiveness which will increase the risk of supply disruption of minerals and resulting price hike."

Strategy to Counter Challenges

"Collaborative research is needed to bridge the existing gap between the production of strategic minerals and their use in various products and to explore the possibilities of devising specific material substitution strategies."

"Research on metal recycling as well as the substitution of strategically important mineral raw materials provides new solutions and value-added core processes in extractive industries."

"The strategic mineral resources can be secured apart from stockpiling, by the way of (i) forming cartels for high-risk items, (ii) investing in overseas mines and modifying regulations to encourage investments in foreign developments, (iii) commodity-based exchange agreements with the nations especially of low-income countries, and (iv) forming strategic alliances and short- and long-term agreements with the nations having abundance of these resources securing trade routes."

Summary and Conclusions

"Strategic minerals play a key role in modern technology development as well as find an indispensable position in the national defense sector, space research, high technology, and electronic industries."

"India has a shortage of resources of some strategic minerals, whereas others such as V, Li, and Co are either scarce or absent."

"The shortfall of strategic raw materials may drastically affect Indian mineral industry and adversely influence economic growth unless urgent steps are taken to bridge this demand–supply gap."

"Dependence of India on a few nations for supplies of the minerals that are crucial for the manufacturing of commercial and defense materials has raised serious concern in the government policies in recent years."

"To meet current economic and developmental goals, India will need to focus on the domestic exploration of strategic minerals and also need to secure these resources through the strategic acquisition of overseas mines and signing diplomatic and trade agreements."

2.2.2 Rare Earth Element Recovery Potentials from End-of-Life Hybrid Electric Vehicle Components in 2010–2030 [13]

This is a machine-generated summary of:

Yano, Junya; Muroi, Takanori; Sakai, Shin-ichi: Rare earth element recovery potentials from end-of-life hybrid electric vehicle components in 2010–2030 [13].

Published in: Journal of Material Cycles and Waste Management (2015).

Link to original: https://doi.org/10.1007/s10163-015-0360-4.

If you want to cite the papers, please refer to the original.

For technical reasons we could not place the page where the original quote is coming from.

Abstract-Summary

"The purpose of this study is to clarify the present and future recovery potential of REEs that are disposed of as EoL HEVs."

"This study first estimated the numbers of EoL HEVs during fiscal years (FYs) 2010–2030, and then clarified the potential for recovery of REEs from two HEV-specific components—the hybrid transmission and NiMH battery unit."

"As of FY2030, REE recovery potentials will increase to 220 tons and 2900 tons for EoL hybrid transmission and NiMH battery units, respectively."

"The potential for recovery of REEs from EoL hybrid transmissions and NiMH battery units is estimated to equal 35.4 and 92.1% of respective demand."

Introduction

"Seo and others [14] estimated the Dy demand for products, including vehicles, in Japan until 2030, and showed that Dy demand for driving motors in HEVs would account for 56% of 1700 t-Dy, the total demand in 2030."

"They showed that a HEV contained a total of 3.5 kg REEs in NiMH (nickel metal hydride) battery and 0.6 kg in hybrid transmission (driving motor and generator), with only 0.4 kg in the remaining components."

"Both forecasting future demand and clarifying REE recovery potentials from EoL HEVs are useful for considering future ELV management."

"REE recovery potential was defined as the amounts of REEs contained in EoL HEV-specific components: the maximum amounts potentially recovered if all EoL components were successfully collected and 100% of REEs recovered from the components."

Materials and Methods

"The lifespan of NiMH battery units is shorter than that of HEVs, and therefore the units must be replaced several times during the vehicle's lifetime, such as during automobile inspections."

"An internet questionnaire survey of HEV owners was used to estimate the replacement probability for NiMH battery units."

"It was assumed that NiMH batteries could be replaced a maximum of two times during the HEVs lifespan."

"The replacement probability was assumed to be the same for all generations of HEVs using NiMH batteries."

"The dismantling surveys conducted in November 2012 collected a NiMH unit used in first-generation HEVs produced in 1997 [15]."

"Using the number of EoL HEVs and NiMH battery units, and the REE contents of a hybrid transmission and a NiMH battery unit, the annual amounts of REEs from EoL HEVs during FY2010–2030 were estimated as recovery potential from the downstream automobile flow."

Results and Discussion

"REE recovery potential equaled 35.4% of the REE demand for new HEV transmissions as of FY2030 if all EoL HEV hybrid transmissions were collected."

"REE recovery potential increased more rapidly than demand because EoL battery units were generated from not only EoL HEVs but also by replacement during the HEV lifespan."

"REE recovery potential equaled 92.1% of REE demand for NiMH battery units in new HEVs as of FY2030 if all EoL NiMH battery units were collected."

"It is also to be noted that REE demand and recovery potentials from other types of next-generation vehicles were not considered in this study."

"The collection efficiency within the EoL HEV management stream, and recovery efficiency during the recycling processes, should also be considered when estimating the recovery amounts of REEs."

Conclusion

"This study first estimated the numbers of EoL HEVs during FY2010–2030, and then clarified the potential for recovery of REEs from two HEV-specific components—the hybrid transmission and the NiMH battery unit."

"Motor magnets in the hybrid transmission and NiMH battery cells were considered as the parts containing REEs."

"REE recovery potentials will increase from FY2010 to FY2030, from 5.0 tons to 220 tons for EoL hybrid transmission, and from 73 to 2900 tons for NiMH battery units."

"As of FY2030, the potential for recovery of REEs from EoL hybrid transmissions and NiMH battery units is estimated to equal 35.4 and 92.1% of respective demand."

2.2.3 Recovery of Cobalt and Lithium Values from Discarded Li-Ion Batteries [16]

This is a machine-generated summary of:

Vishvakarma, Shubham; Dhawan, Nikhil: Recovery of Cobalt and Lithium Values from Discarded Li-Ion Batteries [16].

Published in: Journal of Sustainable Metallurgy (2019).

Link to original: https://doi.org/10.1007/s40831-018-00208-4.

If you want to cite the papers, please refer to the original.

For technical reasons we could not place the page where the original quote is coming from.

Abstract-Summary

"The active material was carbothermally reduced in a muffle furnace at atmospheric conditions for a predetermined time (5–45 min) and temperature (500–900 °C) and the reduced mass was leached in distilled water for 30 min followed by wet magnetic separation."

"That the magnetic fraction contains almost 90% cobalt values, non-magnetic majorly containing graphite and other impurities such as Mn, Cu, and Al and precipitated salt contains lithium carbonate."

"This study demonstrates a simple, environment friendly, and flexible process which includes indigenous carbothermal reduction of the active material, selective dissolution of lithium carbonate, and subsequently wet magnetic separation to recover cobalt values in a short time."

Introduction

"LIBs have a limited lifespan and contain flammable components and valuable metals such as aluminum, iron, copper, lithium (~ 2 to 7%), cobalt (~ 10 to 20%), nickel–manganese along with metallic shells and plastics [17, 18]."

"The five components of LIB are anode (copper plate coated with graphite), a cathode (aluminum plate coated with cathode material (lithium cobalt oxide, lithium nickel oxide, lithium manganese oxide), electrolyte (high-grade lithium salts in organic solvents), separator, and casing."

"It has been reported that lithium carbonate and cobalt can be recovered from LIBs using protective atmosphere such as oxygen-free and reductive vacuum roasting followed by wet magnetic separation [19–21]."

"The key aspect of this work includes indigenous heat treatment of the active material, selective dissolution of lithium carbonate in distilled water, and subsequently wet magnetic separation to recover cobalt and lithium values in a short time."

Materials and Methods

"The grinding followed by sieving of LIBs separated the material into two fractions + 53 μm mostly consist of copper–aluminum foil separator, plastics, and steel casing while − 53 μm fraction mostly consists of the active material."

"The organic binder material was burned off during heating of active material at 450 °C for 30 min and minus 53 μm fraction mostly consists of black powder."

"Since the underlying reduction of LCO is achieved using graphite present in the active material; therefore, the heat treatment is also referred to as carbothermal reduction."

"The material in slurry form was slowly introduced within the magnetic field matrix to ensure there is a clean separation of magnetic values from the non-magnetic fraction."

"The obtained fractions (magnetic, non-magnetic, and leach precipitated salts) were also characterized by XRD, SEM–EDS, and VSM to evaluate phases, composition analysis, and saturation magnetization, respectively."

Results and Discussion

"Significant M_s value (~30 emu/g) was observed for 700 °C sample due to the formation of magnetic fraction (Co and CoO) along with initiation of carbothermal reduction."

"For 900 °C, it was found that with an increase in time duration (5–15 min) there is a gradual increase in saturation magnetization; however, on further increase (15–45 min) there is minimal change in M_s value (~ 10 emu/g) which are in conformance with the XRD findings indicating that the reaction is complete in 15 min at 900 °C."

"As per XRD and VSM results, the best condition corresponding to 900 °C and 15 min was subjected to leaching followed by wet magnetic separation."

"The XRD and EDS analysis revealed that the precipitated salt, non-magnetic, and magnetic fractions consisted of lithium carbonate, graphite along with traces of copper and aluminum and cobalt."

Conclusion

"A carbothermal reduction-leaching-wet magnetic separation process is adopted without using an external additive to extract lithium and cobalt from spent LIB."

"That the magnetic fraction contains almost ~ 90% cobalt values, non-magnetic majorly containing graphite and other impurities such as Mn, Cu, and Al and precipitated salt contains lithium carbonate."

"The processing of 100 g batteries yielded almost 6 g precipitated salt, 17 g non-magnetic, and 13 g magnetic fractions."

"This study demonstrates an environment friendly process which includes indigenous heat treatment of active material, selective dissolution of lithium carbonate, and subsequently wet magnetic separation to recover cobalt values in a very short time."

2.2.4 Life Cycle Assessment and Economic Analysis of Acidic Leaching and Baking Routes for the Production of Cobalt Oxalate from Spent Lithium-Ion Batteries [22]

This is a machine-generated summary of:

Anwani, Sandeep; Methekar, Ravi; Ramadesigan, Venkatasailanathan: Life cycle assessment and economic analysis of acidic leaching and baking routes for the production of cobalt oxalate from spent lithium-ion batteries [22].

Published in: Journal of Material Cycles and Waste Management (2020).

Link to original: https://doi.org/10.1007/s10163-020-01095-2.

If you want to cite the papers, please refer to the original

For technical reasons we could not place the page where the original quote is coming from

Abstract-Summary

"This paper explores two routes of recycling process (acidic leaching and baking) and compares their environmental and economic impacts along with the extraction efficiency and purity of cobalt oxalate."

"The paper uses Box–Behnken method to optimize operating conditions of these two routes, and polynomial equation based on the experimental data of both the routes are developed for the extraction efficiency of cobalt oxalate."

"The optimum extraction efficiency and purity of the cobalt oxalate using acidic leaching route are found as 85.40 and 89.80%, whereas for the baking route, these values stand at 93.87 and 99.20%, respectively."

"Acidic leaching is found to be a greener route with an economic advantage over the acidic baking route."

"The acidic baking route may be used if we desire to have high purity cobalt oxalate, irrespective of its weaker economics and harsher environmental impact."

Introduction

"If discarded into the landfills, these metals will be wasted, and hence recycling of spent LIBs is an essential step to safeguard the environment as well as to have a sustainable supply of these metals to the corresponding industries."

"Ebrahimzade and others [23] have used combination of particle swarm optimization (PSO) and ANN to predict the leaching rate of cobalt from spent LIBs."

"The objective of the economic analysis of the recycling of spent LIBs is two-fold."

"It also compares various processes of spent battery recycling."

"Spent LIBs recycling industry will also be facing the problem of emissions and sustainability."

"There are very few literature which report the environmental assessment of the recycling of spent LIBs."

"Boyden and others [24] compared different recycling processes for recycling of spent LIBs with landfilling in terms of environmental impacts, calculated by GaBi software."

Materials and Methods

"After the preparation of UCM, two parallel routes (acidic leaching and baking) for extraction of cobalt oxalate were carried out, and a comparison of these two processes is presented."

"In the acidic leaching route, seven independent variables (A = leaching temperature, B = solid/liquid ratio, C = leaching time, D = acid molarity, E = H_2O_2 concentration, F = stirring time and G = moles of oxalic acid) and one dependent variable (Y_1 = extraction efficiency of cobalt oxalate) have been chosen."

"In the baking process, three independent variables (H = baking temperature, I = baking time and J = concentrated H_2SO_4 volume) and one dependent variable (Y_2 = extraction efficiency of cobalt oxalate) were defined."

"For water leaching process, three independent variables (K = pulp density, L = water leaching temperature and M = water leaching time), and one dependent variable (Y_3 = extraction efficiency of cobalt oxalate) were used."

Results and Discussion

"These equations will be useful to find the extraction efficiency under a similar environment with the different operating conditions of cobalt oxalate extraction using acidic baking route."

"That, the recovery efficiency and purity of cobalt oxalate in case of acid baking was higher than the acid leaching route."

"The above discussion reveals that baking of UCM, drying and cobalt oxalate extraction have significant environmental impacts on all environmental indices of acidic baking route in a descending order."

"Acidic leaching seems to be a relatively greener route compared to acid baking for cobalt oxalate extraction from spent LIBs."

"It can be concluded that recycling of cobalt oxalate from spent LIBs using either the acidic leaching or the baking route is economically feasible compared to the cost of virgin cobalt oxalate analytical grade material."

Conclusions

"acidic leaching and baking to recover cobalt oxalate have been studied."

"The optimum extraction efficiency of cobalt oxalate was found to be 85.40 and 93.87% with the acidic leaching and baking routes, respectively."

"The maximum purity of cobalt oxalate was found to be 89.80 and 99.20% through the acidic leaching and baking routes, respectively."

"From these indices, it is clear that the acidic leaching route is relatively greener and economical than the acidic baking route."

"The acidic baking route may only be employed for obtaining high purity material, irrespective of its weaker economics and harsher environmental impact."

2.2.5 Estimation of Lithium-Ion Battery Scrap Generation from Electric Vehicles in Brazil [25]

This is a machine-generated summary of:

Cabral-Neto, João Pinto; de Mendonça Pimentel, Rejane Magalhães; Santos, Simone Machado; Silva, Maísa Mendonça: Estimation of lithium-ion battery scrap generation from electric vehicles in Brazil [25]

Published in: Environmental Science and Pollution Research (2022).

Link to original: https://doi.org/10.1007/s11356-022-23730-1.

If you want to cite the papers, please refer to the original.

For technical reasons we could not place the page where the original quote is coming from.

Abstract-Summary

"Among the diversity of electronic waste, lithium-ion batteries (LIB), specifically those used in the propulsion of electric vehicles (EV), are considered pollutants of significant impact."

"This study presents a model for estimating the production of electric vehicles and the generation of scrap LIBs, based on time series, combining battery life, car sales data, and the mileage profile covered by a car in Brazil."

"Around 700 thousand EVs are expected to be circulating in Brazil by 2030, with approximately 500 thousand LIBs to be converted into scrap by 2040."

"The delaying effect of the scrap generated from LIBs is highlighted, in line with the battery life, which, in the future, may have a very negative impact on waste management."

Introduction

"The chemical compounds in the LIB and its ability to release charge also affect the concentrations and types of toxic gases released, and the use of LIBs in EVs is one of the forms of application that most require a battery charging capacity."

"Although lithium is not used just in battery production, about 46% of the world's production is used to manufacture batteries for the most diverse purposes, among them, electric vehicles."

"The National Solid Waste Policy (NSWP) — Law 12,305/2010 (Brazil 22) — deals generically with batteries without distinguishing the technologies, and requires manufacturers, importers, distributors, and resellers to structure and implement systems of reverse logistics of batteries after the end of the product's useful life."

"It aims to contribute to solid waste management since, according to Cabral-Neto and others [26], few studies deal with LIB scrap generation, especially in developing countries such as Brazil."

Material and Methods

"According to Neto and others [27], vehicle sales data compose a non-stationary series that presents both trend and seasonality behaviors."

"Three exponential smoothing models were analyzed, considering the amount of EV, which provided the best adjustments for the data: the Holt–Winters multiplicative model, the integrated autoregressive model of moving averages — ARIMA (double exponential smoothing) — and the simple moving averages model."

"The ARIMA model has two smoothing parameters: α (level) and β (trend), while the moving averages only produce the seasonal adjustment for the forecast."

"The cycle of a lithium-ion battery converted into traveled mileage, according to data in the technical sheet of one of the most popular EV models (Bolt EV), and according to GM [28], corresponds to approximately 416 km/cycle."

Results and Discussion

"It is estimated that the average useful life of a lithium-ion battery is 12 years for the Brazilian traffic situation."

"Combining the LIB's useful life with the projections of the number of EVs circulating on the market until 2030, obtained with the ARIMA model, the number of scrap batteries generated was estimated."

"Due to the 12 years of battery life, the generation of scrap behaves like a rebound effect on vehicle sales."

"The different behavior of lead-acid batteries, with a useful life of approximately 3 years, causes cars to change batteries more often in a shorter time, leading to a cascade effect in the generation of waste (Neto and others [27])."

"The lack of a guideline dealing explicitly with lithium-ion batteries leads to a lack of planning and favors poor management and inappropriate disposal of LIB scrap."

Conclusion

"Using a mathematical model of time series, combined with EV sales, LIB service life, and mileage profiles, estimating the amount of lithium-ion battery scrap is possible."

"Through the statistical adjustment of the precision coefficients, the results indicate that the time series adhere to the prediction processes for both vehicle sales data and LIB scrap projections."

"Forecasts of the annual quantity of EVs sold (2020–2030), as well as the annual forecast of LIB scrap (2020–2040), indicated that the number of scrap batteries generated is entirely out of step with vehicle sales data, proving the delayed effect on the generation of these residues (rebound effect)."

"As LIBs have their life span determined by the number of cycles and since these cycles correspond to a certain duration of the vehicle's autonomy, it is expected that, depending on each country and region's mileage profile, these batteries will last longer or less time."

Bibliography

1. Randive K, Jawadand S (2019) Strategic minerals in India: present status and future challenges. Miner Econ. https://doi.org/10.1007/s13563-019-00189-0
2. Weston R (1984) Strategic materials. Rowman & Allanheld, Philadelphia, p 1
3. Kessel KA (1990) Strategic minerals. U.S. Alternatives, National Defence University Press, Washington, DC, pp 13–14
4. Simandl GJ, Akram C, Paradis S (2015) Selected critical materials: uses, markets and resources. In: Symposium on critical and strategic materials. Symposium on critical and strategic materials, British Columbia Geological Survey Paper 2015-3, 13–14 Nov 2015, pp 1–4
5. Jawadand S, Jain PK, Sharma G, Randive K (2017) Strategic minerals: resources, requirement and challenges for India (Abstract). MEAI international conference and expo on mining industry: vision 2030 and beyond, Nagpur, p 325
6. Planning Commission (2007) Report of the working group on minerals exploration and development (other than Coal and Lignite) for eleventh five year plan (2007–12), January 2007. Government of India. http://planningcommission.nic.in/aboutus/committee/wrkgrp11/wg11_rpmin.pdf
7. Pandey BD (2011) Rare metals extraction from non-ferrous resources in India: present status and prospects of R&D. In: Mankhand T, Majumdar R, Abhilash S (eds) 15th international conference on non-ferrous metals, Kolkata, 8–9 July 2011. https://eprints.nmlindia.org/4266/1Theme_-_B_D_Pandey.pdf
8. Gupta V, Biswas T, Ganesan K (2016) Critical non-fuel mineral resources for India's manufacturing sector, a vision for 2030. In: Council on Energy, Environment and Water (CEEW) and National Science and Technology Management Information System (NSTMIS), Department of Science and Technology, Govt. of India, CEEW Report July 2016, p 104
9. Randive KR, Jawadand SA, Raut TS (2017) National mineral policy and its impact on Indian mineral sector. J Geosci Res Special 1:51–56
10. Speirs J, Gross R, Contestabile M, Candelise C, Houari Y, Gross B (2014) Materials availability for low carbon technologies: an assessment of the evidence. In: A report by the UK Energy Research Centre (UKERC) Technology & Policy Assessment Function, p 50. http://www.ukerc.ac.uk/asset/34D2BFC5-9C0D-4C07-BA6CD6D15BDE549A/
11. Khanna A, Ganeshan S (2010) Critical non-fuel minerals security: why India urgently needs to have a policy in place. The Energy and Resources Institute, New Delhi
12. De Ridder M (2013) The geopolitics of mineral resources for renewable energy technologies. The Hague Centre for Strategic Studies, The Hague, p 28p
13. Yano J, Muroi T, Sakai S-I (2015) Rare earth element recovery potentials from end-of-life hybrid electric vehicle components in 2010–2030. J Mater Cycl Waste Manage. https://doi.org/10.1007/s10163-015-0360-4
14. Seo Y, Morimoto S (2014) Comparison of dysprosium security strategies in Japan for 2010–2030. Resour Policy 39:15–20
15. Sakai S, Yano J, Muroi T, Watanabe N (2013) Dismantling of conventional and hybrid vehicles for end-of-life vehicle management. In: Proceedings of ISWA 2013 World Congress, Vienna, Austria
16. Vishvakarma S, Dhawan N (2019) Recovery of cobalt and lithium values from discarded li-ion batteries. J Sustain Metall. https://doi.org/10.1007/s40831-018-00208-4
17. Poyraz AS, Huang J, Cheng S, Bock DC, Wu L, Zhu Y, Marschilok AC, Takeuchi KJ, Takeuchi ES (2016) Effective recycling of manganese oxide cathodes for lithium based batteries. Green Chem 18:3414
18. Takacova Z, Havlik T, Kukurugya F, Orac D (2016) Cobalt and lithium recovery from active mass of spent Li-ion batteries: theoretical and experimental approach. Hydrometallurgy 9:163
19. Hu J, Zhang J, Li H, Chen Y, Wang C (2017) A promising approach for the recovery of high value-added metals from spent lithium-ion batteries. J Power Sources 351:192
20. Chen X, Ma X, Luo C, Zhou T (2017) Recovery of valuable metals from waste cathode materials of spent lithium-ion batteries using mild phosphoric acid. J Hazard Mater 326:77

21. Zeng X, Li J, Shen B (2015) Novel approach to recover cobalt and lithium from spent lithium-ion battery using oxalic acid. J Hazard Mater 295:112
22. Anwani S, Methekar R, Ramadesigan V (2020) Life cycle assessment and economic analysis of acidic leaching and baking routes for the production of cobalt oxalate from spent lithium-ion batteries. J Mater Cycl Waste Manage. https://doi.org/10.1007/s10163-020-01095-2
23. Ebrahimzade H, Khayati GR, Schaffie M (2019) PSO–ANN-based prediction of cobalt leaching rate from waste lithium-ion batteries. J Mater Cycles Waste Manag 22:228–239. https://doi.org/10.1007/s10163-019-00933-2
24. Boyden A, Kie V, Doolan M (2016) The environmental impacts of recycling portable lithium-ion batteries. Procedia CIRP 48:188–193. https://doi.org/10.1016/j.procir.2016.03.100
25. Cabral-Neto JP, de Mendonça Pimentel RM, Santos SM, Silva MM (2022) Estimation of lithium-ion battery scrap generation from electric vehicles in Brazil. Environ Sci Pollut Res. https://doi.org/10.1007/s11356-022-23730-1
26. Cabral-Neto JP, Pimentel RMM, Santos SM, Silva MM (2020) Perspectives and challenges of vehicle electrification: an academic review. Revista Brasileira de Geografia Física 13:2802–2819. https://doi.org/10.26848/rbgf.v13.6.p2802-2819
27. Neto JC, Silva MM, Santos SM (2016) A time series model for estimating the generation of lead acid battery scrap. Clean Technol Environ Policy 18:1931–1943. https://doi.org/10.1007/s10098-016-1121-3
28. General Motors—GM (2020) Bolt EV Technical Sheet. https://www.chevrolet.com.br/carros/bolt-ev. Accessed 13 May 2020
29. Barik SP, Prabaharan G, Kumar L (2017) Leaching and separation of Co and Mn from electrode materials of spent lithium-ion batteries using hydrochloric acid: laboratory and pilot scale study. J Clean Prod 147:37–43. https://doi.org/10.1016/j.jclepro.2017.01.095
30. Brazil (2010) Lei 12.305/2010—National Solid Waste Policy (NSWP). http://www.planalto.gov.br/ccivil_03/_ato2007-2010/2010/lei/l12305.htm . Accessed 12 Dec 2019 (in Portuguese)

Chapter 3
Technology for EVs

3.1 Introduction by Editor

The nature of the technology of the batteries will determine the material that is used for battery manufacturing. In lithium-ion batteries, lithium is used predominantly on the cathode side whereas graphite is used on the anode side. However, R&D is going on worldwide (partly to offset the scarcity of lithium). In the future, some of these technologies will become commercial and will determine the demand for critical minerals in future.

3.1.1 Understanding the Battery Technology

Every battery has two electrodes—a cathode and an anode—between which trillions of charged atoms travel. The cathode largely determines a battery's performance, cost, and environmental footprint. Broadly, all the components of a cell can be classified into two categories: active and inactive. Active components are responsible for storing energy, and they vary between different cell chemistries [1]. On the other hand, inactive components do not store energy but facilitate the flow of current in and out of the cell or are part of the cell packaging. In the case of lithium-ion cells, during the charging processes, the Li-ions from the host anode (−) material ($LiCoO_2$) "de intercalate" and move through the electrolyte and "intercalate" within the graphite layers of cathode (+), during the discharge process, the reverse happens. This results in the continuous movement of Li-ions between two zones during the charging and discharging processes causing the current to flow.

Cathode Material (+) [or LIC Chemistries]: The Li-ion is a very generic term and as these ions can be generated from Cathode using a variety of chemistries, bringing a range of performance characteristics. The most common chemistry combinations used in lithium cobalt oxide (LCO), lithium manganese oxide (LMO), Lithium iron

S. Ray and D. Mukherji (eds.), *A Primer on Electric Vehicles in India*,
https://doi.org/10.1007/978-981-97-8966-5_3

phosphate (LFP), lithium nickel manganese cobalt oxide (NMC), and lithium nickel cobalt aluminum oxide (NCA), lithium titanate oxide (LTO) and cell manufacturers also use their combination to get an improved performance.

Out of the seven basic cell components, five are always constant irrespective of the type of battery. The remaining two are active components; namely, the anode and the cathode active material. The active materials are the only ones that change between chemistries: the separators in the battery provide a barrier between the anode (−) (or negative) and the cathode (+) (or positive) while enabling the smooth exchange of (Lithium) ions from one side to the other.

[1] Sarkar et al. (2018).

3.1.2 Types of Batteries

It is customary to use the name of the active material for classifying the variations in the battery. Almost 99% of commercial Li-ion batteries use graphite as the anode, and thus the cathode remains the only unique differentiator. The five large-scale commercialized types are lithium cobalt oxide (LCO), lithium manganese oxide (LMO), Lithium iron phosphate (LFP), lithium nickel manganese cobalt oxide (NMC), and lithium nickel cobalt aluminum oxide (NCA). All of these are abbreviations for the name of the cathode material used in the batteries [2].

v LCO: The first lithium chemistry to be introduced was LCO, which stands for lithium cobalt oxide. As the name indicates, this cathode material contains three elements, out of which lithium and cobalt are the key materials, obtained from mining. LCO is good for achieving a high energy density (150 Wh/kg), which is advantageous in electronic devices. However, the limited cycle life (lower than 1,000) restricts its application to electronic devices. Another challenge with LCO is its relatively lower temperature tolerance limit (150 °C). This implies that the cells cannot be fast charged or discharged, as this would produce excessive heat internally, leading to safety problems. This is not considered a limitation for electronic devices since a typical cell phone takes 4 h to charge and 24 h to discharge, both of which are classified as very slow cycling. Understandably, LCO batteries have been dominating the electronic device market. Overall, from a safety and longevity point of view, LCO technology is not considered ideal for EV applications.

v LMO: Some of the limitations of LCO were overcome with the introduction of LMO, or lithium manganese oxide. Interestingly, manganese oxide was already well known to the battery industry as the cathode material in zinc-alkaline batteries. LMO not only has a high thermal limit of stability (250 °C), but it also allows fast cycling (3C and above) with low cell internal resistance. Both qualities make it ideal for high-power applications. These advantages coupled with the low cost of LMO make it an attractive cathode material, in spite of its lower cycle life and energy density. The low cost is primarily due to the inexpensiveness of its main constituent, manganese.

v NMC: This cathode material gets its name from its primary constituent elements, lithium nickel manganese cobalt oxide. The variations of NMC such as 111, 311 (or 622), and 811 indicate the relative concentrations of Ni, Mn, and Co. A minor compositional change in the material yields large performance benefits. The cycle life and fast discharge capability are an improvement over LCO without compromising on the energy density. The overall optimum performance of NMC cells on all parameters has ensured that their use is growing steadily for all major battery applications, namely transportation and stationary storage.

v NCA: The latest addition to the Li-ion family is NCA, which stands for lithium nickel cobalt aluminum oxide. The key benefit of NCA is its high energy density (250 Wh/kg). A high energy density is a definite advantage (it translates to a longer driving range) for EVs, which is where NCA cells have found their biggest market. NCA technology is less suitable for high-power applications, where fast charging and discharging are required. When used under high charge rates (c-rates) [3], the cycle life and the thermal resistance of the battery are significantly reduced unless protected by the design and usage of proper materials.

v LFP: Other than LMO, LFP is the only other member of the Li-ion family that is unaffected by the price of cobalt. This is because both LMO and LFP do not contain cobalt. LFP stands for lithium iron phosphate (Ferrum is the chemical name of iron). The phosphate has the unique property of a high decomposition temperature (more than 400 °C), giving it a strong safety advantage. The cycle life and high-power capability of LFP cells are comparable to those of NMC cells, but their energy density is relatively lower (130 Wh/kg). In spite of the lower energy density, LFP cells have found good traction in the transportation sector, particularly for electric buses, which have a higher tolerance for heavier batteries. Other than this, LFP seems suited for stationary applications, where the battery weight is not a critical parameter.

v LTO: (Lithium titanate oxide) This variation of the technology uses a different anode compared to all the previously discussed technologies, which use graphite as the anode. In these cells, the anode is LTO, which stands for lithium titanium oxide. In LTO cells, the cathode may be either LMO, NMC, or LFP. However, as the anode is what differentiates it from the other technologies, this abbreviation is used to denote the technology. The energy density of LTO cells is approximately 30% lower in comparison with the other chemistries mentioned above due to its lower voltage (approximately 2.4 Volts). However, the cycle life of these cells is the highest among all chemistries. Moreover, the cells can be cycled at very high c-rates of 6C–10C without affecting the cycle life or safety and with minimal impact on the RTE. As a result, LTO batteries are ideally suited for EV applications where fast charging is crucial.

[2] https://www.wri.org/research/state-research-development-electric-vehicle-battery-technology

A summary of some of the prominent LIC chemistries are indicated in Table 3.1.

[3] Explained below.

Table 3.1 Summarizes the characteristics of Li-ion batteries

Li-ion chemistries							
Category	Unit	Lithium iron phosphate	Lithium manganese oxide	Lithium titanate	Lithium cobalt oxide	Lithium nickel cobalt aluminium	Lithium nickel manganese cobalt
Cathode (+) Chemistry Descriptor	–	LFP $LiFePO_4$	LMO $LiMn_2O_4$	LTO Li_2TiO_3	LCO $LiCoO_2$	NCA Li $(NiCoAl)O_2$	NMC Li $(NiMnCo)O_2$
Volts/ Cell	V	3.2 ~ [4] 3.3	3.8	2.2 ~ 2.3	3.6 ~ 3.8	3.6	3.6 ~ 3.7
Specific Energy	Wh/ kg	80 ~ 130	105 ~ 120	70	120 ~ 150	80 ~ 220	140 ~ 180
Energy Density	Wh/ L	220 ~ 250	250 ~ 265	130	250 ~ 450	210 ~ 600	325
Cycle Life	–	1000 ~ 2000	> 500	> 4000	> 700	> 1000	1000 ~ 4000
Operating Temperature Range	Deg C	~ 20 to + 60	~ 20 to + 60	~ 40 to + 55	~ 20 to + 60	~ 20 to + 60	~ 20 to + 55
Self-Discharge	%	< 1%	5%	2 ~ 10%	1 ~ 5%	2 ~ 10%	1%

Source Author

The performance and longevity of Li-ion batteries are highly dependent on the following four parameters:

Ø Chemistry of electrodes and electrolyte: The chemical properties of the electrodes and electrolyte fundamentally govern the cycle life of the cell. Once the chemistry is fixed, the operation parameters—temperature, C-rate, and DoD— primarily affect the cycle life.

Ø Temperature: As the ambient temperature around the cell is increased, the capacity fade rate increases and the cycle life decreases. The comfortable temperature range for Li-ion cells is approximately 20–30 °C. Continuous operation at temperatures outside this range can significantly lower the cycle life. An approximate rule of thumb is that the cycle life of cells decreases by a factor of 2× for every 10 °C increase in ambient temperature. Since the daytime temperatures in most parts of India are above 35 °C in summer, the manufacturer needs to evaluate the longevity under conditions that mimic these real-world conditions.

Ø C-rate: At an increased c-rate, a higher current pass through the cell, leading to higher heat generation in the cell due to internal resistance. A sustained high c-rate mimics conditions of a higher operating temperature, resulting in faster capacity fade. Within Li-ion, the cells are classified as power cells or energy cells based on the specific construction design parameters. If a high c-rate is required for the operation, power cells may be more cost-effective even though their cost is higher (in $/kWh) in comparison with energy cells.

Ø Depth of Discharge (DoD): The DoD has a very significant impact on the capacity fade of Li-ion batteries. A reduction in the DoD can prolong the cycle life. In general, the aging of cells is most pronounced at the extremities of the State of Charge (SoC), that is, when it is closest to being fully charged or fully discharged. A common strategy to prolong the cycle life is to have restricted cycling between 10 and 90%, or 20 and 80%, thus avoiding the extremities of SoC.

The cost of each component of the cell value chain is given in Fig. 3.1. The cost of the cathode is the highest and is approximately 51%. The manufacturing cost of a cell is 24% of the total cost and consists of a series of assembly processes. Cell and battery packs consist of a highly material-intensive value chain and need very specialized processes to manufacture.

[4] ~ refers to approximately.

Li-Ion cell value chain break up with cost:

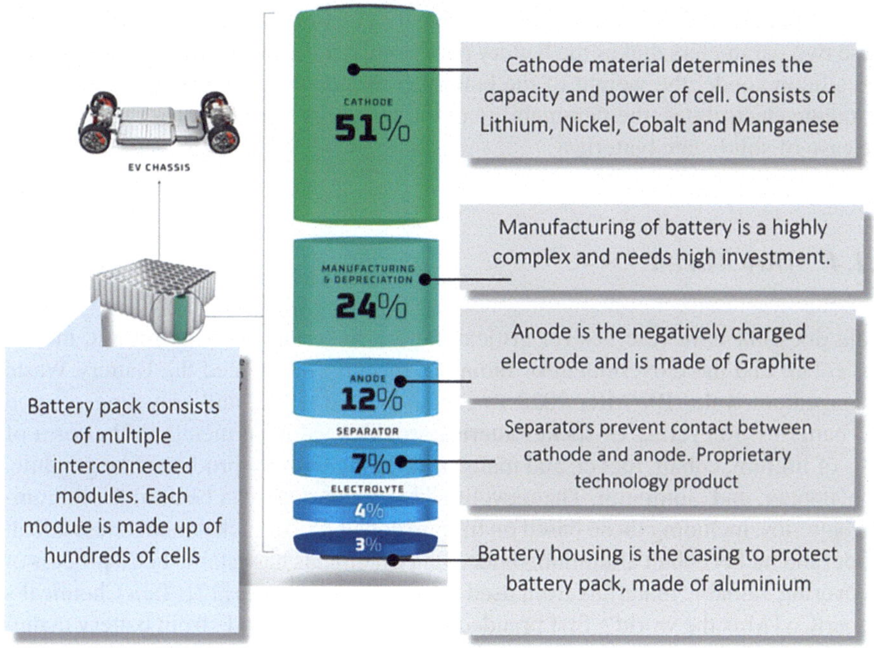

Fig. 3.1 Li-Ion cell Cost stack worldwide. Source: Author

3.1.3 Alternative Technologies

There is a need to indigenize battery manufacturing in India and this is being encouraged through the PLI in ACC. CSIR-CECRI is tying up with IIT Hyderabad to localize lithium-ion battery technology [5, 6]. Also, companies like Amara Raja Batteries have started on developing lithium-ion technology in the country [7, 8]. Recycling metals, when the vehicle is scrapped, is important to recover metals. The other development is the research on alternatives e.g. lithium sulphur, zinc-ion, aluminum air, graphene, etc.

There are some technologies that are on the horizon: [9] these include Lithium-sulphur, silicon anode, solid state batteries etc.

In addition to the above-described Li-ion battery technologies, several others are currently under development. The R&D community and industry are working on various aspects of existing battery technologies and also on futuristic battery technologies with the objectives of lowering the cost and improving performance [10]. One of the main thrust areas of global R&D activities is to improve the volumetric (Wh/L) and gravimetric (Wh/kg) energy density. A combination of improvements in these two parameters makes the battery pack compact. In case of the lithium sulphur and silicon anode, the weight of the battery comes down significantly. With liquid inorganic electrolytes, the flammability of the battery is reduced. This is also true in the case of solid-state batteries.

3.1.4 Way Ahead

India does not have reserves for critical battery materials such as lithium, nickel, and cobalt and the Government of India has recently introduced the Battery Waste Management Rule (BWMR) 2022 to encourage the re-use and recycling of used EV batteries. Recycling of spent batteries provides valuable metals in the form of salts of lithium, cobalt, nickel, and manganese along with by-products like graphite, iron, copper, and aluminum. The recycling [11] technology can be used for lithium-ion batteries, including those based on lithium cobalt oxide, nickel manganese cobalt oxide, and nickel cobalt aluminum oxide. Tata Chemicals has established a process of recovering valuable materials from used Lithium-ion batteries [12]. Tata Chemical's InsperiCoTM is the world's first branded recycled cobalt. Aside from battery manufacturing [13], there are opportunities for efficient battery management software and monitoring battery performance and remaining useful life. Following several recent fire incidents involving EV batteries in India and voluntary recalls by EV producers, there is a growing emphasis on the quality, performance, and safety aspects of EV batteries. A look at the mining, material processing, and battery manufacturing capabilities shows the dominance of China in all the categories. This will be discussed further in Chap. 4.

(This is repeated below) [5] https://www.autocarpro.in/news-national/pure-ev-partners-csircecri-to-localise-lithiumion-battery-tech-77974

[6] Tata is building an EV battery plant in Gujarat. https://www.ndtv.com/india-news/tata-to-build-rs-13-000-crore-ev-battery-plant-in-gujarat-4094455#:~:text=Tata%20To%20Build%20%E2%82%B9%2013%2C000%20Crore%20EV%20Battery%20Plant%20In%20Gujarat&text=Salt%2Dto%2Dsoftware%20conglomerate%20Tata,of%20around%20%E2%82%B9%2013%2C000%20crore.

[7] https://www.business-standard.com/article/companies/amara-raja-starts-work-on-lithium-ion-cells-sets-up-tech-hub-at-tirupati-121022100451_1.html

[8] https://www.bloomberg.com/news/articles/2023-08-01/battery-firm-plans-1-billion-lithium-ion-plant-in-india?utm_medium=social&utm_source=twitter&utm_campaign=socialflow-organic&cmpid=socialflow-twitter-business&utm_content=business

[9] https://www.wri.org/research/state-research-development-electric-vehicle-battery-technology

[10] https://www.adlittle.com/sites/default/files/prism/Achieving%20resilience%20and%20sustainability%20for%20the%20EV%20battery%20supply%20chain.pdf

[11] https://planet.outlookindia.com/news/global-li-ion-battery-recycling-market-to-be-worth-6-9-billion-by-2030-news-415029

[12] https://www.tatachemicals.com/Asia/News-room/Press-release/tata-chemicals-lion-battery-operations#:~:text=Li%2Dion%20battery%20recycling%20recovers,raw%20materials%20from%20the%20earth

Machine Generated

Ding et al. (2019) review the automotive LIBs in terms of market, battery costs, critical element analyses, currently state-of-art Li-ion chemistries, and next-generation battery chemistries. Shende et al. (2021) propose a method that reduces frequent significant peaks in the current input and output from the battery, giving better health of the battery over several charging and discharging cycles and a significant boost to the available mileage of batteries. Banez-Chicarro et al. (2013) analyze the effects of introducing EVs in a power system and compare different charging strategies. Assad et al. (2017) argue that the use of Electric Vehicles (EVs) as dynamic electrical energy storage systems in the smart grid offers numerous advantages while affecting the grid and EV battery pack. Gupta and Shankar (2022) suggest a way so that the construction of infrastructure to integrate energy storage in utility grids does not remain a barrier to the adoption of new technologies in grids due to costs. Saidharan et al. (2022) examine the way in which managing the usage of renewable energy sources along with battery storage can optimize the operations of grid-connected EV charging stations. Verma et al. (2022) develop a methodology to help a planner choose a number of charging points of different technologies in the distribution network without adversely affecting its reliability, stability, and performance. Sarita8 et al. (2013) analyse the FCEV roadmap taking into account manufacturing and market developments, prospects of technologies, products and services contributing to the design of complex innovation value chains ranging from technology to market entrance of FCEVs and building strategies for linking FCEVs development with other related industries (suppliers and consumers of related products/technologies).

References

Autocar (2022) Pure EV partners CSIR-CECRI to localise lithium-ion battery tech. Link: https://www.autocarpro.in/news-national/pure-ev-partners-csircecri-to-localise-lithiumion-battery-tech-77974

Business Standard (2021) Amara Raja starts work on lithium-ion cells, sets up tech hub at Tirupati. Link: https://www.business-standard.com/article/companies/amara-raja-starts-work-on-lithium-ion-cells-sets-up-tech-hub-at-tirupati-121022100451_1.html

Dutz, P., Hoffman, F., Schreiber, B., Seidel, P., Krug, A., Navarro, R., Zander, K. O., Akayama, S. () Prism: Achieving Resilience and Sustainability for the EV Battery Supply Chain. Arthur D. Little report.

Maitra, B., Schlosser, A., Sempf, F., Doerr, K.M., Chaudhari, P., Yadav, S., Manjari, R. (2023) Toward a Sustainable Value Chain for Electric vehicles In India. Arthur D. Little report.

NDTV (2023) Tata To Build ₹ 13,000 Crore EV Battery Plant In Gujarat. June 5. Link: https://www.ndtv.com/india-news/tata-to-build-rs-13-000-crore-ev-battery-plant-in-gujarat-4094455#:~:text=Tata%20To%20Build%20%E2%82%B9%2013%2C000%20Crore%20EV%20Battery%20Plant%20In%20Gujarat&text=Salt%2Dto%2Dsoftware%20conglomerate%20Tata,of%20around%20%E2%82%B9%2013%2C000%20crore.

Outlook Planet Desk (2023) Global Li-Ion Battery Recycling Market To Be Worth $6.9 Bn By 2030. April 11. Link: https://planet.outlookindia.com/news/global-li-ion-battery-recycling-market-to-be-worth-6-9-billion-by-2030-news-415029

Sarkar, E. M., Sarkar, T., and Bharadwaj, M. D. (2018) Lithium-ion battery supply chain: enabling national electric vehicle and renewables targets. Current Science, Vol. 114, No. 12 (25 June 2018), pp. 2453–2458.

Tata Chemicals (n.d.) Press release Tata Chemicals launches li-ion battery recycling operations. Link: https://www.tatachemicals.com/Asia/News-room/Press-release/tata-chemicals-lion-battery-operations#:~:text=Li%2Dion%20battery%20recycling%20recovers,raw%20materials%20from%20the%20earth.

WRI (2022) State of Research & Development in Electric Vehicle Battery Technology. Link: https://www.wri.org/research/state-research-development-electric-vehicle-battery-technology

Parthiyal, S. (2023) Battery Firm Plans $1 Billion Lithium-Ion Plant in India. Bloomberg. August 1. Link: https://www.bloomberg.com/news/articles/2023-08-01/battery-firm-plans-1-billion-lithium-ion-plant-in-india?utm_medium=social&utm_source=twitter&utm_campaign=socialflow-organic&cmpid=socialflow-twitter-business&utm_content=business

Footnotes/Endnotes

[1] Sarkar et al. (2018).

[2] https://www.wri.org/research/state-research-development-electric-vehicle-battery-technology

[3] Explained below.

[4] ~ refers to approximately.

[5] https://www.autocarpro.in/news-national/pure-ev-partners-csircecri-to-localise-lithiumion-battery-tech-77974

[6] Tata is building an EV battery plant in Gujarat. https://www.ndtv.com/india-news/tata-to-build-rs-13-000-crore-ev-battery-plant-in-gujarat-4094455#:~:text=Tata%20To%20Build%20%E2%82%B9%2013%2C000%20Crore%20EV%20Battery%20Plant%20In%20Gujarat&text=Salt%2Dto%2Dsoftware%20conglomerate%20Tata,of%20around%20%E2%82%B9%2013%2C000%20crore.

[7] https://www.business-standard.com/article/companies/amara-raja-starts-work-on-lithium-ion-cells-sets-up-tech-hub-at-tirupati-121022100451_1.html

[8] https://www.bloomberg.com/news/articles/2023-08-01/battery-firm-plans-1-billion-lithium-ion-plant-in-india?utm_medium=social&utm_source=twitter&utm_campaign=socialflow-organic&cmpid=socialflow-twitter-business&utm_content=business

[9] https://www.wri.org/research/state-research-development-electric-vehicle-battery-technology

[10] https://www.adlittle.com/sites/default/files/prism/Achieving%20resilience%20and%20sustainability%20for%20the%20EV%20battery%20supply%20chain.pdf

[11] https://planet.outlookindia.com/news/global-li-ion-battery-recycling-market-to-be-worth-6-9-billion-by-2030-news-415029

[12] https://www.tatachemicals.com/Asia/News-room/Press-release/tata-chemicals-lion-battery-operations#:~:text=Li%2Dion%20battery%20recycling%20recovers,raw%20materials%20from%20the%20earth

[13] Maitra et al. (2023).

3.2 Machine Generated Summaries

Disclaimer: The summaries in this chapter were generated from Springer Nature publications using extractive AI auto-summarization: An extraction-based summarizer aims to identify the most important sentences of a text using an algorithm and uses those original sentences to create the auto-summary (unlike generative AI). As the constituted sentences are machine selected, they may not fully reflect the body of the work, so we strongly advise that the original content is read and cited. The auto generated summaries were curated by the editor to meet Springer Nature publication standards. To cite this content, please refer to the original papers.

Machine generated keywords: storage, energy storage, smart, grid, storage system, charge, bms, device, system, utility, charge station, profile, station, smart grid, management system.

3.2.1 Automotive Li-Ion Batteries: Current Status and Future Perspectives [1]

This is a machine-generated summary of:

Ding, Yuanli; Cano, Zachary P.; Yu, Aiping; Lu, Jun; Chen, Zhongwei: Automotive Li-Ion Batteries: Current Status and Future Perspectives [1].

Published in: Electrochemical Energy Reviews (2019).

Link to original: https://doi.org/10.1007/s41918-018-0022-z.

If you want to cite the papers, please refer to the original.

For technical reasons we could not place the page where the original quote is coming from.

Abstract-Summary

"Lithium-ion batteries (LIBs) are currently the most suitable energy storage device for powering electric vehicles (EVs) owing to their attractive properties including high energy efficiency, lack of memory effect, long cycle life, high energy density and high power density."

"Modern EVs, however, still suffer from performance barriers (range, charging rate, lifetime, etc.) and technological barriers (high cost, safety, reliability, etc.), limiting their widespread adoption."

"Novel battery chemistries and technologies including high-energy electrode materials and all-solid-state batteries are also evaluated for their potential capabilities in next-generation long-range EVs."

Introduction

"To mitigate these problems, developing green, sustainable energy together with alternative fuel vehicles, particularly battery-powered vehicles, has become a promising alternative in the automotive industry."

"With the invention of gasoline-powered vehicles in 1910 [2], electric automobiles were gradually displaced from the market due to their relatively high costs, limited range (40 miles) and speed (45 miles per hour)."

"Besides the range and costs, some other crucial factors that influence the extensive market penetration of electric automobiles require further improvement including battery lifetime (at least 10 years), safety, and reliability, and charging infrastructure."

"Considering theoretical energy limitations of current Li-ion chemistries (lithium transition oxide cathodes and graphite anodes: 350 Wh kg^{-1} at the cell level) for automotive industries [3], new high-energy and high-capacity battery materials including Ni-rich NMC cathode materials, silicon-based anode materials, and innovative battery chemistries including all-solid-state batteries (SSBs) are evaluated and discussed for next-generation EV applications in the short and long terms."

Markets, Costs and Critical Material Analyses for Automotive Batteries

"With the continuous increase in production capacity of batteries, battery costs will be further reduced, which is very favorable for future market penetration of EVs."

"Several critical raw materials (e.g., lithium, cobalt) should be evaluated through the use of a scenario-based supply and demand analysis owing to their limited resources (lithium reserves: 12–14 million tons (Mts)-based Li metal equivalent; cobalt reserves: 7.1 Mts), restricted extracting technology and high demand from growing automotive industries and large-scale grid energy storage [4]."

"To predict lithium demand in the short term (2016–2025) and in the long term (2016–2035), lithium consumption and supply are roughly calculated by two scenarios: the low-growth scenario (15% CAGR) and the high growth scenario (30% CAGR), since the CAGR of total battery production capacity (GWh) was between 15 and 30% during 2011–2016 [5]."

The State-of-the-Art Li-Ion Chemistries for Automotive Batteries

"Compared with LCO, other cathode materials including spinel LMO, NCA, NMC and LFP have become the most prominent battery chemistries for EVs because they not only possess more stable crystal structures, but also have much more abundant resources and lower costs, thus showing large application potential for EVs."

"Compared to spinel LMO, NCA is more expensive, but has advantages of higher specific capacity (200 mAh g^{-1} at the material level), high energy density (> 200 Wh kg^{-1} at the cell level) and long life (calendar life: > 15 years)."

"The severe voltage fading is probably ascribed to the transition toward a spinel phase when the material is cycled at a cutoff window of 2.0 and 4.6 V. Besides cycling performance, low electronic conductivities and low tap densities need to be enhanced before HE-NMC could be considered as a potential battery technology for next-generation EV applications [6]."

Next-Generation Li-Based Automotive Batteries

"The poor cycling performance of pure NMC-811 is attributed to a structural transformation at the particle surface owing to the high reactivity of Ni^{4+} with electrolytes, leading to the increase in charge-transfer resistance between the cathode and the electrolyte during repeated cycling."

"Since energy density of a battery is directly proportional to its cell voltage, high-voltage cathode materials should be developed for matching the high potential of the LTO anode, such as $LiN_{i0.5}Mn_{1.5}O_4$ (4.7 V) [7]."

"Li metal anodes possess a very high theoretical specific capacity (3860 mAh g^{-1}) and the lowest electrochemical potential (− 3.04 V vs. the standard hydrogen electrode, SHE), both of which are very promising for future high-energy LIB application."

"When replacing liquid electrolytes with SSEs while keeping the same anode material (graphite), the volumetric energy density will not be increased while the gravimetric energy density will decrease by 10% owing to the high density of solid materials compared with liquid materials."

Conclusions and Outlook

"This review article gives a comprehensive analysis and discussion for automotive LIBs in terms of market, battery costs, critical element analyses, currently state-of-art Li-ion chemistries and next-generation battery chemistries."

"In the period between 2016 and 2025 (short term), EVs are still required to depend on the relatively mature battery technologies including NMC, NCA and LMO–NMC-blended chemistries for developing new battery chemistries with high specific energy and lower costs such as Ni-rich NMC (e.g., NMC-811) and Si."

"Considering the theoretical specific energy limitation (350–400 Wh kg^{-1}) of such battery systems, it is highly imperative for developing new higher energy battery chemistries in the long term (beyond 2025) such as SSBs (e.g., NMC-811/SSE/Li, LNMO/SSE/Li), Li–S, Li–air batteries."

"Besides the concerns of energy density and costs for LIBs, element resources of battery materials, especially for critical elements such as Li and Co, have a significant effect on the market penetration of electric automobiles."

3.2.2 Sizing Scheme of Hybrid Energy Storage System for Electric Vehicle [8]

This is a machine-generated summary of:

Shende, Vijayendra; Singh, Krishna Veer; Bansal, Hari Om; Singh, Dheerendra: Sizing Scheme of Hybrid Energy Storage System for Electric Vehicle [8].

Published in: Iranian Journal of Science and Technology, Transactions of Electrical Engineering (2021).

Link to original: https://doi.org/10.1007/s40998-021-00416-x.

If you want to cite the papers, please refer to the original.

For technical reasons we could not place the page where the original quote is coming from.

Abstract-Summary

"Energy storage system (batteries) plays a vital role in the adoption of electric vehicles (EVs)."

"Li-ion batteries have high energy storage-to-volume ratio, but still, it should not be charged/discharged for short periods frequently as it results in degradation of their state of health (SoH)."

"UCs provide the advantage of quick and frequent charging/discharging without degrading the battery SoH and are also used to absorb most of the energy generated due to regenerative braking."

"The major source of energy is Li-ion cells which provide the energy required to run the vehicle, whereas the UCs will provide above-average energy required by the motor."

"A power-based method to find the specifications of UCs and batteries is described which provides specifications for ESS hybridization."

"The proposed method, which is based on the prescribed set of limiting values of current and voltages, tries to maintain the UC voltage and battery current within range."

"This method reduces the above-average peaks of the required current from the batteries."

"While recharging due to the regenerative braking, the proposed method removes the above-average peaks of the charging current of UCs."

"The proposed topology along with the EMS provides better state of charge (SoC) levels, giving a 38.6% increase in SoH levels of the batteries."

Introduction

"These Li-ion batteries get charged quickly, discharged as and when required, deliver high energy to storage ratio, and can satisfy substantial power requirements."

"When this converter is connected before all the parallel-connected storages, voltage fluctuations of UCs will be same as that of battery; hence, this scheme (parallel SA–HES or PSA-HES) is not a very good improvisation over the P-HES."

"In capacitor CFA-HES (CCFA-HES) topology, the batteries are decoupled from the load with a DC-DC converter."

"Parallel FA-HES (PFA-HES) is another topology in which all the storage devices are decoupled from the load using separate DC-DC converters."

"Increasing requirement of storage voltage in EVs requires creating standard battery and UC modules."

"The proposed strategy provides a method for making high-voltage storages using low voltage battery and UC modules, i.e., pack level hybridization."

Proposed Methodology

"Assuming almost constant voltage to the load with the help of EMS, the model will consider the load power requirement/supply as a direct function of current passing through the battery."

"These modules will be capable of supplying the current through parallel combination while simultaneously providing the given voltage due to series connections, hence supplying enough energy for a driving cycle."

"It means that 3 individual battery modules, each of 48 V, in series combination would be enough to provide 140 V. The power requirement and the threshold limits will also be linearly multiplied by 3."

"In this design, the EMS assumes that the batteries are the central power supply unit of the vehicle, and UC modules are the main power absorbers for the regenerative braking of the vehicle."

"While regenerative braking, the UC module will absorb the current till the 200A, after which the battery module will absorb the surplus."

Results and Analysis

"The use of UC modules is relevant for the proposed method as the aim is to decrease the usage of batteries at high current supply or absorption rates."

"In the case of not using the capacitor and no limits on the discharge and charge current to the battery, it is evident that the battery module is supplying an excess of 400A to the load and absorbing around 200A while regenerative braking scenarios."

"Reconsidering the facts that were taken care of when designing the battery module, a current limit of 292A is applied to the discharge current of the battery, while the current required above the 292A mark is supplied by the capacitor module."

"A limit of 200A is applied to the charging current of the capacitor, while the excess of which will be readily absorbed by the battery module."

"Further, it enables controlled current discharge from the batteries as well as the capacitors, maintaining similar voltages and temperatures across the storage."

Conclusion

"The current limiting strategy works to limit the voltage fluctuations and power drawn from the batteries."

"The batteries are used as a primary source of power, while ultra-capacitors are used to provide and absorb peak power requirements."

"The proposed method reduces frequent significant peaks in the current input and output from the battery, which has been quantitatively proven by higher SoC levels using this scheme than conventional methods."

"The proposed method also provides better health of battery over several charging and discharging cycles, giving a significant boost to the available mileage of batteries."

3.2.3 IoT Enabled Monitoring of an Optimized Electric Vehicle's Battery System [9]

This is a machine-generated summary of:

Asaad, Mohammad; Ahmad, Furkan; Alam, Mohammad Saad; Rafat, Yasser: IoT Enabled Monitoring of an Optimized Electric Vehicle's Battery System [9].

Published in: Mobile Networks and Applications (2017).

Link to original: https://doi.org/10.1007/s11036-017-0957-z.

If you want to cite the papers, please refer to the original.

For technical reasons we could not place the page where the original quote is coming from.

Abstract-Summary

"For the development and improvement of smart grid, Internet of Things (IoT) technology is an important enabler."

"Use of Electric Vehicles (EVs) as dynamic electrical energy storage system in smart grid offers numerous advantages while affecting the grid and EV battery pack."

"To diminish the impact of mass adoption of EVs, this paper proposes an optimization model aimed at maximizing the trade revenue for EV' aggregator for enabling the smart charging."

"Further, to circumvent the possibility of damage to the EV battery, real-time Battery Monitoring System (BMS) using less complex and easy to implement Enhanced Coulomb Counting Method for SoC estimation and messaging based MQTT as the communication protocol is presented."

Introduction

"The integration of Smart Grid tools in utility networks will influence marvelous transformation in grid management and electric power use in upcoming years."

"Increased deployment of energy storage devices in the distribution grid will help expedite this process and improve system performance [10]."

"Bulk energy storage has been used for decades in the utility grid, and now the integration of renewables is creating a need for more distributed storage [11]."

"To prolong the life of battery-based energy storage system and ensure their reliability, a proper battery monitoring system needs to be integrated along with."

"Since the inception of the smart grid concept being an integral part of IoT was put forward, it has been vastly appreciated by utilities."

"IoT integrated Smart Grid results in improvement of energy productivity, reduced ecological influence, enhanced security, reduced vulnerability to external interference and increased consistency of electric supply [12]."

Electric Vehicles and Smart Grid

"The smart grid is the intelligent integration of generation, transmission, distribution to efficiently deliver sustainable, secure and economical electrical energy."

"In parked condition, EVs generate or store electricity which can be fed back to the grid using appropriate connections- this is known as vehicle-to-grid power or V2G power."

"The batteries of EVs plugged into charging infrastructure can act as distributed energy storage systems for the electrical grid."

"The electrical energy delivered backed to the grid must be priced such that the additional cost incurred must be recovered back as the battery's life cycle is reduced due to frequent charging-discharging."

"Compared to centralized power generation, the comparatively lesser investment costs of vehicle power systems and the small incremental expense to alter EVs to generate grid power imply economic affordability. [13] suggests that EVs possibly won't produce bulk power, because the unit cost of energy from EVs is greater than bulk electricity from centralized generators and due to their elemental engineering features."

Smart Charging

"Optimal scheduling of EVs charging is a necessity to counter the increasing burden on the grid."

"Continuous monitoring of power system is necessary while charging a large number of EVs, to achieve grid balancing."

"Low peak hours during the night, wherein the load is at its minimum, are most suitable for home based charging, using either Level 1 or Level 2."

"Public charging stations (CSs), where most of the charging is expected to occur during day time, can shift their charging load to low peak hours to enhance the functioning of the grid."

"This can be done by deploying attractive charging schemes for the consumers during low peak hours or by storing energy during this period as a backup for peak load hours."

"For simplistic benchmark method, the forecast function is given as: Where $y_{(t)}$ represents the demand in period t, j is the forecast lead time ($j \leq s_2$)."

Battery Monitoring System

"Battery Monitoring System (BMS) is a smart system whose function is to monitor the vigor of a battery pack."

"Using uninterrupted observing each cell of battery pack, BMS pre-identifies the decline in the cell performance."

"It also generates audio and visual warnings by the minimum and maximum values of key variables, e.g., discharges current, operating temperature, etc., pre-set by the consumer as per his need or when a fault is triggered while the battery cells are coupled to the battery monitoring system."

"These vital parameters are also logged and communicated to users for better management/upkeep of battery pack."

"BMS is unable to sense movable connections present in the battery, leakage of cell material, corrosion of connections leading to the development of high resistance and subsequently fire danger."

SoC Estimation

"Being a vital parameter depicting battery performance, precise estimation prolongs the battery lifecycle, avoids deep discharges and helps designing practical control methods to keep battery operating in the optimum region."

"By estimating the present capacity of the battery, it can be safely charged/discharged at levels appropriate for battery lifecycle enhancement."

"The energy capacity of a battery depends upon its charging current, discharging current, oldness, operating temperature, cut-off voltage, and usage profile."

"Numerous techniques have been suggested for the batteries SoC estimation [14, 15]."

"operating temperature, battery usage history, discharge current, and cycle life [16]."

"Few inefficiencies are there in this technique- the initial SoC value is not precisely known, the presence of self-discharge phenomena can change the real SoC value after a prolonged storage time, accurate measurement of current going through the battery, temperature dependency and battery degradation due to aging should be taken into consideration."

Hardware Implementation

"voltage, current flow and temperature by appropriate sensing technology and interfaced through an Arduino Uno used as ADC device to microcontroller Broadcom BCM2837 SoC based Raspberry Pi 3 Model B. The SoC estimation technique has been implemented by measuring voltage, current and temperature of two parallel connected 3S Li-Po (Lithium Polymer) batteries with total nominal capacity of 4400 mAH."

"As MQTT delinks the publisher and subscriber, only the information about hostname/IP and port of the broker is sufficient to communicate with messages."

"There are three layers of security in MQTT [17]: 1) Network level: The communication between broker and clients can be secured by using physically secured network or Virtual Private Network (VPN) as a foundation."

"The battery packs electrical parameters are measured using hall effect current/ voltage sensors and after amplification, are communicated with the processor Raspberry Pi 3 B interfaced through an ADC."

"The coulomb counting algorithm is implemented in the Node-RED environment and estimated SoC is communicated via MQTT protocol."

Results

"The working of SoC estimation technique and real time of communication of battery parameters were tested while discharging the LiPo batteries."

"The external battery temperature measured is communicated in real time."

"The parameters calculated and communicated to users are the State of Charge, State of Health and Battery Temperature."

Conclusion

"The implementation of Smart Grid devices in the utility grid will influence vast modification in grid management and usage of electric power in upcoming years."

"The detailed structure of smart grid is discussed in this work."

"With Electric Vehicles (EVs) being an integral part of the sustainable development, mass unplanned deployment will affect the grid in a negative way."

"The impact of EVs on the grid is discussed and to mitigate this impact by facilitating smart charging, a complete optimization model is presented aimed at maximizing the trade revenue for an aggregator of EV."

3.2.4 Smart Charging Profiles for Electric Vehicles [18]

This is a machine-generated summary of:

Banez-Chicharro, Fernando; Latorre, Jesus M.; Ramos, Andres: Smart charging profiles for electric vehicles [18].

Published in: Computational Management Science (2013).

Link to original: https://doi.org/10.1007/s10287-013-0180-8.

If you want to cite the papers, please refer to the original.

For technical reasons we could not place the page where the original quote is coming from.

Abstract-Summary

"The charging pattern of EVs is the main factor that determines these effects."

"Unregulated charging (probably when returning home) would have undesirable consequences (e.g. increase in variable costs, emissions, reduction of reliability) for the system, it is therefore necessary to develop an "intelligent" charging strategy."

"These characteristics justify the existence of different smart charging profiles."

"This document compares different possible strategies for charging EVs and their consequences in the power system."

"The Spanish power system for 2020 is analysed under different EV penetration levels and charging strategies."

"The results show the benefits of using smart charging profiles instead of an unregulated profile, obtaining large cost reductions and maintaining system reliability levels."

"The benefits of using a day-ahead management system are also evaluated, resulting in a small reduction of system variable cost compared to the use of pre-defined charging profiles."

Introduction

"Unregulated charging, which consists on the users charging their EVs as they return home, will involve an increase of the peak demand, increasing system variable costs while decreasing its reliability (Bañez Chicharro and others [19])."

"In Bañez Chicharro and others [19], an evaluation of the economic and environmental impacts of EV deployment for the power systems of Spain, Portugal and Greece under different pre-defined charging strategies is performed."

"The authors of Kiviluoma and Meibom [20] analyse two different EV charging methods (unregulated and smart with V2G capability) and the potential influence on the operation costs of an optimised power system."

"The authors developed a statistical locational marginal price and wholesale energy cost model instead of an operational model to evaluate EV impacts."

"This paper analyses the impacts of EVs on the power system under different charging strategies (unregulated and three smart strategies) and a wide range of different penetration levels."

Model Description

"The ROM model follows a combined modelling approach that replicates the sequence of the planning and real-time operation of power systems under a System Operator (SO) point of view."

"Hourly profiles for demand, wind (generation and forecasted errors), other RES generation and operational reserves are introduced in the model as input data."

"In combination with the technical and economic characteristics of thermal and hydro units, the model calculates the UC and ED of the system in order to supply the demand and the operating-reserve requirements minimising the operational costs of the system, which includes thermal costs and cost of not-served energy (NSE)."

"To introduce EVs in the optimisation model, some data about the technical characteristics of the EV, and on the mobility patterns of the commuters are necessary: distance travelled, hourly usage and connection profiles, energy storage of the batteries, minimum and maximum state-of-charge (SOC), specific EV consumption, and efficiencies (grid-to-battery, battery-to-wheel)."

Charging Strategies for Electric Vehicles

"The EV data are introduced as an input to the valley-filling model, which provides the EV charge profile that is then introduced as an input to the ROM model together with system data."

"The maximisation of the minimum total demand and the flattening of the demand curve has been added as sub-objectives to avoid the random distribution of the EV charge when the WG spillage minimisation is fulfilled."

"The RES integration strategy calculates the operation of the power system without introducing EVs, and obtains the WG spillage that would occur."

"This WG spillage is then introduced along with the EV data as inputs in the RESi model to obtain the EV charge strategy that minimises wind spillage."

"The charging strategy is introduced into the ROM model together with other system data to obtain the economic and operation results of the system with EVs."

Case Study and Scenarios

"Twelve different EV penetration degrees have been analysed, ranging between 0%, which represents the base case scenario, and 50% EVs, which corresponds to 15 million EVs."

"Regarding EV charging, four different EV charging profiles have been considered."

"EV operational management is analysed to determine whether it would be worthwhile to have a system that is able to communicate and manage EV charge in day-ahead schedule or if it would be enough to have predefined EV charge profiles."

"Two alternative scenarios are considered: day-ahead and pre-defined profiles."

"In order to evaluate the impact of using different EV charging profiles, these profiles have been analysed under the twelve EV penetration degrees with day-ahead management."

"In order to assess the convenience of day-ahead management, the VF and RESi charging profiles have been analysed under the twelve EV penetration degrees with both day-ahead management and pre-defined profiles."

Results

"With lower penetration of EVs, the COST strategy supplies more energy with coal and RES, resulting therefore in less energy supplied with CCGTs and GTs, than the other two smart profiles."

"The COST strategy charges between 15 and 19 h while the VF and RESi charge between 15 and 21 h. Furthermore, contrary to what VF and RESi profiles do, the COST profile increases EV demand during the evening hours, as the volume of EVs in the system increases."

"For the VF strategy, using a pre-defined profile instead of a day-ahead one makes the average fuel cost to increase as the number of EVs in the system does."

"Day-ahead profiles tend to use more energy from coal units that when a pre-defined profile is adopted to charge the EVs."

"The day-ahead profile charges more energy during the night hours (using more energy from coal units) than the pre-defined profile, which charges more during the evening (producing more with GTs and pumped-storage hydro units)."

Conclusions

"This paper analyses the effects of introducing EVs in a power system and compares different charging strategies."

"The importance of using "intelligent" strategies to charge EVs is evaluated by comparing smart strategies with an unregulated EV charging."

"Different objectives for the smart charging of the EVs have been considered: cost minimisation, valley-filling approach and RES, namely wind, integration."

"Regarding power system operation, the three smart charging strategies have a very similar effect, which differ from the use of unregulated charging."

"Smart charging profiles require lower volumes of energy production to supply the same demand, and the use of pumped-storage hydro units is reduced, while they also result in higher production of coal generating units."

"The smart charging strategies presented in this paper have been designed to fulfil different objectives."

3.2.5 Adoption of Battery Management System in Utility Grid: An Empirical Study Using Structural Equation Modeling [21]

This is a machine-generated summary of:

Gupta, Laxmi; Shankar, Ravi: Adoption of Battery Management System in Utility Grid: An Empirical Study Using Structural Equation Modeling [21].

Published in: Global Journal of Flexible Systems Management (2022).

Link to original: https://doi.org/10.1007/s40171-022-00319-8.

If you want to cite the papers, please refer to the original.

For technical reasons we could not place the page where the original quote is coming from.

Abstract-Summary

"Understanding the adoption of battery management systems (BMS) or energy storage systems (ESS) is essential for utilities interested in developing efficient grid systems."

"This research enhances the understanding of ESS adoption and its success rate in grid utility."

"This research addresses the concerns regarding which factors are essential for ESS adoption in smart grid and how they impact the adoption rate."

"This study describes empirical testing of a structural model based on the TOE (technology–organization–environment) theory of adoption influencing the adoption rate of the battery management system."

"Based on the best knowledge of the authors, this is probably the first attempt that explicitly explains a model for the factors affecting ESS integration in the grid for developing countries like India and focused on battery adoption."

Introduction

"It is essential to examine when and how energy storage systems in smart grid technologies would be adopted by utility companies, which supply most of the power consumed by consumers and industrial users."

"Utilities face significant operational challenges when incorporating energy storage systems with smart grid technology, and adoption rates are still low."

"This study bridges the gap, identifies the issues, and determines which factors are driving or deterring utilities from adopting energy storage systems in the smart grid to upgrade India's power infrastructure."

"The objective of this research is as follows: RO1: What factors may affect utility companies' interest and ability to adopt battery management or energy storage system in the grid."

"This research explores factors that influence the adoption of battery management systems with smart grid technologies based on interviews with experts from Indian electric utilities."

Literature Review

"With the urgent need to improve performance and sustainability of ESS in these systems, development of growing techniques and operation in electric grid applications have been deployed from energy storage perspectives (Liu and others [22])."

"Because of the lack of well-developed theory in the smart grid domain, there is a need for research that addresses utility companies' issues while deploying energy storage."

"With increasing electrical demand on an overloaded grid and a need to improve the efficiency of the energy supply, today's utilities industries face unprecedented challenges (Tenggren and others [23])."

"With increase in the use of energy storage technologies shares in the electric grid, the power sector seeks to create large-scale manufacturing facilities for indigenous storage devices."

"It is critical to identify the difficulties and establish which variables are motivating or discouraging utilities from adopting energy storage systems in the smart grid to upgrade India's power infrastructure and lessen its environmental implications."

Research Model and Hypothesis

"According to the literature and theory studied, the first research hypothesis is expressed as follows: Perceived Social Trust significantly affects adoption rate of ESS/ BMS in smart grid."

"Consumer's attitude toward the deployment of smart grid technologies is an important factor in the adoption of ESS within the smart grid."

"The success of ESS adoption may be attributed to public perception of the technology, which leads to the following hypothesis: Public attitude toward ESS technology significantly affects adoption rate of ESS/ BMS in smart grid."

"We presume that knowledge and awareness about technology have positive impact on adoption rate to implement ESS in grid: Knowledge/Awareness about technology significantly affects adoption rate of ESS/ BMS in smart grid."

"The seven elements used in this study that determine the rate of adoption are: Perceived social trust, Perceived benefits of innovation, Public attitude towards ESS technology, Awareness/Knowledge about technology, Perceived risk/Comptetive pressure, Cost comptetion and Environmental awareness."

Research Methodology

"The software calculates and analyzes linear structural relationships between different latent constructs (factors) in their recursive or non-recursive path models."

"Stage 1 Using the TOE framework, the research assistant retrieves and organizes questionnaire forms that examine the factors significantly influencing the adoption rate of ESS in the utility grid."

"A CFA includes the estimate of measurement model, in which the observed variables are mapped onto the latent constructs based on theory and prior testing."

"The relationship between the variables, as well as the questionnaire questions (indicators) that operationalize the evaluation of features, are all included in this model."

"Using structural equation modeling, researchers can determine how effectively these indicators (observed variables) represent latent components (unobserved variables)."

"Equation model (SEM) examines latent variables (factors) for causal relationships and statistically defines the hypothesized structure."

"The assessment of a path model through SEM is similar to the evaluation of a measurement model through CFA."

Results

"A multicollinearity test is used to determine whether a correlation exists between the independent variables in the regression model."

"No factor explains more than 50% of the total variance in the dataset, hence these results satisfy Harman's one-factor test, proving that CMB is not an issue in the collected data."

"The goal of convergent validity testing is to determine whether or not the items of a latent variable (factor) can effectively represent their corresponding component."

"The values of factor correlation coefficients are compared with square root of the AVE to test discriminant validity."

"To determine discriminant validity, we check if the square root of each latent construct's AVE value is significantly bigger than any correlation between latent constructs."

"For good discriminant validity of the measured model, the square root of AVE (i.e., given in italics) for each element is much bigger than the correlation coefficients with other variables (Fornell & Larcker [24])."

Discussion and Findings

"This implies that the factors perceived social trust (PST), perceived benefits of innovation (PBI), Public attitude toward ESS technology (PAT), knowledge and awareness about technology (KAT), Cost competition (CC), and environmental awareness (EA) contribute significantly to the adoption rate (ADR) of an energy storage system in the grid."

"It will have a significant impact on the adoption of particular technologies, which will affect the impact of the SG implementation at the planning stage itself. (6) Perceived social trust has a positive significant effect on the ESS adoption rate, implying that trust and social confidence are essential factors to consider when implementing or deploying any new technology, hence it must be addressed for achieving sustainable energy systems with wider adoption of smart grid technologies. (7) Lastly, perceived risk/competitive pressure has a negative impact on grid adoption of ESS technology."

Theoretical and Managerial Implications

"This research established the possible success factors for battery storage adoption at the power grid from a theoretical standpoint."

"Although there have been numerous studies on smart grid adoption technologies, this research has primarily concentrated on using energy storage systems or battery management systems in utility grids."

"This research advanced our knowledge and understanding of energy storage system adoption in grid utility and its success rate."

"Considering managerial perspective, our findings suggest that for adopting ESS in the utility grid, the electricity service provider should focus mainly on the knowledge and awareness about energy storage technology, cost competition, and perceived benefits in the market related to storage technologies."

"This paper will help policymakers in many sectors of the economy, from conventional utilities to businesses and customers who want to use storage technology in power grids."

Conclusion and Policy Implications

"The results show that the factors perceived social trust, the public attitude toward ESS technology, cost competition, perceived benefits of innovation, environmental awareness, awareness/knowledge about technology positively influence the adoption of ESS in the grid, whereas perceived risk (competitive pressure) influence negatively on the adoption of ESS."

"A key takeaway for power sector policymakers is that they should invest more in innovative activities such as government, universities, and schools may educate the general public and students about energy storage technologies and the benefits of future electricity systems and markets."

"Cost reduction in grid storage technology is one of the most critical variables influencing storage technology adoption across all models."

"For this different international and national agency should lead the charge to allocate an appropriate budget for the construction of infrastructure to integrate energy storage in utility grids so that cost will not remain a barrier to the adoption of new technologies in grids."

3.2.6 Comparative Analysis of Deep Learning Models for Electric Vehicle Charging Load Forecasting [25]

This is a machine-generated summary of:

P Sasidharan, Manujith; Kinattingal, Sundareswaran; Simon, S. P.: Comparative Analysis of Deep Learning Models for Electric Vehicle Charging Load Forecasting [25].

Published in: Journal of The Institution of Engineers (India): Series B (2022).

Link to original: https://doi.org/10.1007/s40031-022-00798-4.

If you want to cite the papers, please refer to the original.

For technical reasons we could not place the page where the original quote is coming from.

Abstract-Summary

"Grid-connected plug-in electric vehicle charging stations having integrated renewable energy sources like photovoltaic (PV) systems with battery energy storage help manage the variability of electric vehicle charging load and reduce the stress on the grid."

"Such charging stations will benefit from short to medium term forecasting of load demand, as it allows them to optimize the operation through better utilization of the PV and battery stored energy rather than relying on the grid."

“Some of the widely researched deep learning models such as long short-term memory (LSTM), gated recurrent units (GRU), hybrid of convolution neural network (CNN) and LSTM, hybrid of CNN and GRU, multivariate LSTM and multivariate GRU are analyzed for fitment for the charging load forecasting problem.”

“The datasets available from multiple charging stations in a region are used for training the models.”

Introduction

“EV characteristics such as charging time of the batteries, driving range, and maximum load capacity of the vehicle also play key roles in shaping the EV load profile.”

“The above attributes must be factored when a forecasting model is built for EV charging stations.”

“An operational strategy to utilize energy sources giving preference to PV first and then to battery storage over the grid for EV charging is explained in the literature [26].”

“EV charging load forecasting has multiple challenges associated with it.”

“With the advancement in the machine learning techniques and the availability of large datasets on the EV charging load from different parts of the world, it has become far more efficient to rely on advanced machine learning techniques to forecasting the EV load.”

“Training is performed on these models using the EV charging data spanning multiple years gathered from public charging stations in the region under study.”

Methods and Techniques

“A model with multivariate input fed into LSTM network is considered in this study for comparison with models with single variable input.”

“The first stage has the 1D Convnet which takes the sequence as input, extracts the features, does down sampling of data using Maxpooling layer and passes the output to LSTM model.”

“In this option, multiple 1 D CNN layers and pooling layer are used at the first stage of the model and the same extracts the patterns in the input sequence data.”

“In LSTM model with single variable, the temporal distribution of load sequence data is fed into the model for prediction of load at the next time step or a future time step.”

“In Model 5, the multivariate input signals are passed on to an LSTM layer with 64 units and having activation function as ReLU.”

Results and Discussion

“The real charging power values for 72 h are fed to the models.”

"The popular ML models were used to predict the EV load values, as a base for evaluating the DL models."

"The deep learning models (Model 1 through Model 6) are subsequently taken up for training, prediction and performance evaluation."

"The models are tested by predicting the EV load values for 24, 48 and 72 h in future."

"The predictions made using Model 6 also yielded similar results."

"The fact that the plots of the predicted values in multivariate deep learning models had close resemblance with those of the actual data emphasizes the superior performance of these models in comparison with the other models."

Conclusion

"The EV charging load forecasting is an important task in managing the usage of renewable energy sources along with battery storage and optimize the operations of grid connected EV charging stations."

"The authors have analyzed six different deep neural network models comprising of LSTM and GRUs, suitable for effective forecasting of time series data, for prediction of EV charging load."

"The real EV charging data available from the charging stations in a region in US was used for training the models."

3.2.7 A Novel Methodology for the Planning of Charging Infrastructure in the Scenario of High EV Penetration [27]

This is a machine-generated summary of:

Verma, Mandhir Kumar; Mukherjee, Vivekananda; Yadav, Vinod Kumar; Ghosh, Santosh: A novel methodology for the planning of charging infrastructure in the scenario of high EV penetration [27].

Published in: Soft Computing (2022).

Link to original: https://doi.org/10.1007/s00500-022-07622-7.

If you want to cite the papers, please refer to the original.

For technical reasons we could not place the page where the original quote is coming from.

Abstract-Summary

"This article presents a novel methodology for distribution network expansion planning (DNEP) considering the inclusion of electric vehicles (EVs), especially, electric bus (EB) charging loads."

"The proposed methodology provides coordinated planning for the development of EB charging station infrastructure that takes into account the effects of both the power dispersion framework and transportation framework."

"The effectiveness of the proposed methodology is investigated by applying it to the 69-node IEEE modified distribution test system considering three charging technologies, viz."

"The results of the proposed model are compared with the direct statistical method, and it revealed that the right selection of technology for EB charging and the right planning of the charging infrastructure can effectively optimize the cost of EV charging infrastructure and thereby catalyze the decarbonization of the transportation sector."

Introduction

"An innovative, dynamic evaluation and optimization technique was employed for siting and sizing of the host charging station (HCS), based on the response surface methodology (RSM) introduced by Box and Wilson 21."

"RSM has also been applied in the field of EV for the optimization of charging infrastructure design (Bayram and others [28]) in the past."

"A methodology for quantifying the optimal dispatch and charging of EBs is proposed considering three technologies, viz."

"The placement and the sizing of HCS, along with the number of charging ports (CPs) at each HCS, are considered to be the objective functions of the proposed methodology. (b) Comparative analysis of the multiple near-optimal solutions is carried out against a set of criteria, including power losses, voltage deviations, and setup expenses. (c) RSM is employed in the present work for obtaining the most optimal solutions from a set of near-optimal solutions."

Problem Formulation

"The EB battery charging process is considered to be continuous, i.e., without any voltage fluctuation and power cut, until the charging state of up to 85% capacity is reached."

"It is also considered that every EB will get seven hours in the night for a full normal charge, and it will leave PCS with full SOC."

"Considering one swapping per hour, every HCS must have a minimum of 8 spare charged batteries, and further, if the number of ports is greater than 5, then the number of spare batteries must be equal to the thrice number of ports available at that particular HCS."

"To obtain optimal values of system losses, power flow, thermal loading, and optimal EB charging, the following constraints are considered in the present work. (a) The energy-storing system constraints are described as follows."

[Section 3]

"It is considered that the location of the HCS on each route should be such that the EBs reaching the HCS shall have SOC of 10 to 20%, and 80 to 85% while leaving."

"This is so because the EBs must reach the HCS before complete drain-out and will leave the HCS before 100% SOC as the last 20% of charging is slow and require more time as compared to the time taken to reach 20% to 80% SOC."

Methodology

"A 69-node IEEE modified distribution test system (Savier and Das [29]) is considered to evaluate the effect of EBs' charging on power quality and reliability."

"If the total charging ports to be installed at an HCS are 6 (BST-, FCT-, and UFCT-based ports), the second row (no."

"The charging time is considered as 5, 15, and 120 min for BST, UFCT, and FCT, respectively (An and others [30])."

"The time which is considered in the present work is the time that the EB will consume in the swapping process. (d) Battery life (BL): Battery life is considered 'long' for normal charging."

"As per the guidelines of the Ministry of Power, GOI (GoI-MoP 2018), for normal charging (as well as for battery swapping), 15 kW (model-Bharat DC-001) charger and, for fast charging, 50 kW (CHAdeMO) charger is considered for each charging port."

The Optimization Model: RSM

"Response surface methodology (RSM) is an optimization technique used to acquire the exact models of the design parameters in terms of response factors in the wake of recognizing their relative commitment, utilizing the Taguchi technique."

"It can take on a wide assortment of practical structures, so it will regularly function admirably as a guess to the genuine response surface. (b) Utilizing the technique for least squares, it is easy to evaluate the parameters in the second-order model."

"As per Taguchi's approach, for the mathematical modeling of the second-order RSM methodology, design parameters are considered as: the number of charging points (NCP), the investment required for setting up HCS (HCSI), energy tariffs (TR), and plug-in period (PIP)."

Results and Discussion

"The following subsections present the results of the above analysis, i.e., the calculated results of the impact of design parameters (NCP, HCSI, TR, and PIP) on output responses (PR, TSL, and BL) of the optimization process with RSM and, finally, the results obtained from RSM are verified with those of the calculated results."

"After NCP reaches a value of 12, the variation in PR increases exponentially because, for a higher value of NCP, the FCT and the UFCT ports dominate over BST ports."

"It is evident from the curve that there is an exponential increase in PR with the increase in PIP since, as compared to BST ports, an increase in FCT and UFCT ports will sharply increase PR and also will take more plug-in time due to their charging characteristics."

"With changing configurations, the number of BST ports increases (hence the number of the FCT and UFCT ports decrease), while the PR decreases."

Conclusion

"The proposed methodology will help the planner prioritize and integrate a significant number of charging points of different technologies into the distribution network without adversely affecting its reliability, stability, and performance."

"As a scope of future work, the proposed model can be further improved by considering the following. (a) Recompute a fresh power assignment profile during the day itself, considering the personal and commercial EV charging at night. (b) In highly polluted cities like New Delhi, India, to reduce the impact on the environment further, the power generation for charging stations may be from renewable energy sources. (c) The spare capacity of the HCS can be utilized for private and commercial EV charging. (d) The temperature variation factor has not been considered in the present work, which can be included in future work."

3.2.8 A Transition Management Roadmap for Fuel Cell Electric Vehicles (FCEVs) [31]

This is a machine-generated summary of:

Saritas, Ozcan; Meissner, Dirk; Sokolov, Alexander: A Transition Management Roadmap for Fuel Cell Electric Vehicles (FCEVs) [31].

Published in: Journal of the Knowledge Economy (2018).

Link to original: https://doi.org/10.1007/s13132-018-0523-3.

If you want to cite the papers, please refer to the original.

For technical reasons we could not place the page where the original quote is coming from.

Abstract-Summary

"With the technological development, recent discussions about global warming and climate change bring public support for emission free vehicles."

"Despite of the advancements and support, the speed of introduction of FCEVs is still not at the desirable levels."

"From a transition management perspective, the present paper seeks to answer the underlying factors behind the implementation of the FCEVs."

"Following a discussion the key drivers of change for the FCEV sector and wild cards with disruptive effects, the paper proposes a strategic roadmap template to set an agenda for a successful transition towards FCEVs."

Introduction

"The discussion in the present paper focuses on how to achieve transition towards fuel cell electric vehicles (FCEVs), which have been considered as a future vision for the automotive industry."

"Market roadmaps make assumptions of customers' behavior, take into account competing products etc. but neglect the more or less 'hidden' stakeholders, which become obvious if one analyses the systemic impact of FCEVs as innovations within industry value chains in relation to the transportation sector in a broader sense."

"The roadmap presented in the study, entitled 'FCEV Global Market Creation', undertakes a broader analysis of Society, Technology, Economy, Environment and Politics (STEEP) at two stages."

"The second section of the paper provides a background for the study and motivations and main issues for the transition into FCEVs."

"Following, a systematic and holistic analysis of external drivers and stakeholders will be presented in the fourth section to uncover broader set of issues affecting the FCEVs."

Background and Objectives

"Among the alternatives for future energy sources, hydrogen comes as one of the first options."

"Hydrogen can be used to power vehicles as well as to be used in fuel cells to generate and store energy."

"Both of these technologies can be used separately or on the same platform, where a car can be powered by a hydrogen combustion engine and the fuel cell as a source of energy to supply electric power in place of a conventional alternator."

"A number of refuelling stations are required for hydrogen-powered cars."

"Adequate refuelling infrastructure will be required for the FCEVs to be marketed as a credible and attractive alternative to conventional vehicles."

"Once hydrogen refuelling stations (HRSs) are available, the initial uptake of the FCEVs will be limited by the cost of buying and using the vehicles."

Approach and Methodology

"The process begins with a wider scanning and surveying phase (Miles and others [32]), where the implications of key trends and drivers of change are examined with implications for the future FCEV markets."

"A typical scanning work involves an analysis of trends, drivers of change, weak signals, wild cards and discontinuities (Saritas and Miles [33])."

"Identification of 'external drivers' through the analysis of societal, technological, economic, environmental and political determinants of FCEVs 2."

"Analysis and mapping of stakeholders to identify key players in each STEEP category The external drivers (STEEP) are described in brief with emerging key issues under each of them."

"The roadmap template provided at the end of the present paper is expected to be filled by the policy and strategy makers promoting the development of the FCEV technologies."

Characteristics of External Drivers

"Some of the technological factors associated to FCEVs are as follows: Reliability of supply Equipment (hardware) Vehicle technologies The supply of hydrogen is an important issue for the creation of FCEV markets."

"As one of the key drivers for hydrogen cars, economic factors in this domain involve a number of points to be considered, including the following: Cost of FCEVs Demand Investments Fuel cells (FCs) are still very expensive, even when compared to BEVs."

"In order to overcome the commercialisation challenge, a close cooperation is needed between vehicle manufacturers; equipment manufacturers in production of fuel cells, hydrogen refuelling stations and hydrogen technology components and sub-systems; fuel retailers; hydrogen producers; energy utilities; and the government departments such as science, technology and innovation, transport and energy."

Characteristics of Stakeholders

"Developing effective roadmaps hence requires knowledge and information about actual but also potential stakeholders, which follow their own strategies and might have the potential to block technologies and innovation diffusion or at least create obstacle and barriers to delay diffusion."

" $\rightarrow$ low, $\nearrow$ medium, $\uparrow$ high Overall, the individual stakeholders have only moderate influence on innovation diffusion, but associations representing car owners have considerable power at different levels."

"These refer to direct competitors, who aim at similar application but using competing technological solutions and surrounding actors like infrastructure suppliers and fuel producers."

" $\rightarrow$ low, $\nearrow$ medium, $\uparrow$ high The analysis shows that competing technological solution providers have potentially stronger influence on the diffusion of innovation."

"National level policy makers will have reasonable influence to design measures, which are supportive to FCEVs diffusion, still the implementation of any federal measures is at the regional and the municipalities' level."

Development of an Eagle Eye FCEV Roadmap

"A roadmap for the creation of a market for FCEVs is a tool for long-term complex planning, which allows setting strategic goals and estimating potential contribution of new technologies, products and services to build a competitive and sustainable FCEV market."

"The FCEV roadmap presents estimated indicators of economic efficiency of the potentially prospective technologies and products, with the high demand potential and attractive consumer properties of FCEVs within the time scale also taking into account the elaborated stakeholder views."

"The roadmap should provide a detailed analysis of market pull, including the following: Areas of product's application determining the demand for technological solutions."

"The roadmap takes into account manufacturing and market developments, prospects of technologies, products and services contributing to the design of complex innovation value chains ranging from technology to market entrance of FCEVs and allows building strategies for linking FCEVs development with other related industries (suppliers and consumers of related products/technologies)."

Bibliography

1. Ding Y, Cano ZP, Yu A, Lu J, Chen Z (2019) Automotive li-ion batteries: current status and future perspectives. Electrochem Energy Rev. https://doi.org/10.1007/s41918-018-0022-z
2. Bryan FR (2012) The birth of Ford Motor Company, Henry Ford Heritage Association. http://www.hfha.org. Accessed 20 Aug 2012
3. Zhang S, Ueno K, Dokko K et al (2015) Recent advances in electrolytes for lithium–sulfur batteries. Adv Energy Mater 5:1500117
4. Gruber PW, Medina PA, Keoleian GA et al (2011) Global lithium availability. J Ind Ecol 15:760–775
5. Pillot C (2017) The rechargeable battery market and main trends 2016–2025. In: BATTERIES 2017, Nice, France
6. Andre D, Kim SJ, Lamp P et al (2015) Future generations of cathode materials: an automotive industry perspective. J Mater Chem A 3:6709–6732. https://doi.org/10.1039/c5ta00361j
7. Arrebola JC, Caballero A, Cruz M et al (2006) Crystallinity control of a nanostructured LiNi0.5Mn1.5O4 spinel via polymer-assisted synthesis: a method for improving its rate capability and performance in 5 V lithium batteries. Adv Func Mater 16:1904–1912
8. Shende V, Singh KV, Bansal HO, Singh D (2021) Sizing scheme of hybrid energy storage system for electric vehicle. Iranian J Sci Technol Trans Electr Eng. https://doi.org/10.1007/s40998-021-00416-x
9. Asaad M, Ahmad F, Alam MS, Rafat Y (2017) IoT enabled monitoring of an optimized electric vehicle's battery system. Mob Netw Appl. https://doi.org/10.1007/s11036-017-0957-z
10. Roberts BP, Sandberg C (2011) The role of energy storage in development of smart grids. Proc IEEE 99(6):1139–1144
11. Vartanian C (2008) The coming convergence, renewables, smart grid and storage. IEEE Energy 2030
12. Fang X, Misra S, Xue G, Yang D (2012) Smart grid — the new and improved power grid: a survey. IEEE Commun Surv Tutorials 14(4):944–980
13. Kempton W, Tomić J (2005) Vehicle-to-grid power fundamentals: calculating capacity and net revenue. J Power Sources 144(1):268–279
14. Watrin N, Blunier B, Miraoui A (2012) Review of adaptive systems for lithium batteries state-of-charge and state-of-health estimation. In: 2012 I.E. transportation electrification conference and expo, ITEC 2012
15. Piller S, Perrin M, Jossen A (2001) Methods for state-of-charge determination and their applications. J Power Sources 96(1):113–120
16. Ng KS, Moo CS, Chen YP, Hsieh YC (2009) Enhanced coulomb counting method for estimating state-of-charge and state-of-health of lithium-ion batteries. Appl Energy 86(9):1506–1511
17. HiveMQ. MQTT Security Fundamentals (Online). Available: http://www.hivemq.com/blog/introducing-the-mqtt-security-fundamentals. Accessed 15 Aug 2017
18. Banez-Chicharro F, Latorre JM, Ramos A (2013) Smart charging profiles for electric vehicles. Comput Manage Sci. https://doi.org/10.1007/s10287-013-0180-8
19. Bañez Chicharro F, Dietrich K, Ramos A, Latorre JM (2012) Evaluation of the impact that a progressive deployment of EV will provoke on electricity demand, steady state operation, market issues, generation schedules and on the volume of carbon emissions. Subtask VI. Technical Report, Deliverable 3.2 (Part II) of the MERGE EU Project
20. Kiviluoma J, Meibom P (2011) Methodology for modelling plug-in electric vehicles in the power system and cost estimates for a system with either smart or dumb electric vehicles. Energy 36(3):1758–1767
21. Gupta L, Shankar R (2022) Adoption of battery management system in utility grid: an empirical study using structural equation modeling. Glob J Flex Syst Manage. https://doi.org/10.1007/s40171-022-00319-8
22. Liu J, Hu C, Kimber A, Wang Z (2020) Uses, cost-benefit analysis, and markets of energy storage systems for electric grid applications. J Energy Storage 32:101731

23. Tenggren S, Wangel J, Nilsson M, Nykvist B (2016) Transmission transitions: barriers, drivers, and institutional governance implications of Nordic transmission grid development. Energy Res Soc Sci 19:148–157
24. Fornell C, Larcker DF (1981) Evaluating structural equation models with unobservable variables and measurement error. J Market Res 18(1):39–50
25. Manujith PS, Kinattingal S, Simon SP (2022) Comparative analysis of deep learning models for electric vehicle charging load forecasting. J Inst Eng (India): Ser B. https://doi.org/10.1007/s40031-022-00798-4
26. Negarestani S, Fotuhi-Firuzabad M, Rastegar M, Rajabi-Ghahnavieh A (2016) Optimal sizing of storage system in a fast charging station for plug-in hybrid electric vehicles. IEEE Trans Transport Electrif 2(4):443–453
27. Verma MK, Mukherjee V, Yadav VK, Ghosh S (2022) A novel methodology for the planning of charging infrastructure in the scenario of high EV penetration. Soft Comput. https://doi.org/10.1007/s00500-022-07622-7
28. Bayram IS, Michailidis G, Devetsikiotis M, Granelli F (2013) Electric power allocation in a network of fast charging stations. IEEE J Sel Areas Commun 31(7):1235–1246. https://doi.org/10.1109/JSAC.2013.130707
29. Savier JS, Das D (2007) Impact of network reconfiguration on loss allocation of radial distribution systems. IEEE Trans Power Deliv 22(4):2473–2480. https://doi.org/10.1109/TPWRD.2007.905370
30. An K, Jing W, Kim I (2019) Battery-swapping facility planning for electric buses with local charging systems. Int J Sustain Transp 13:1–13. https://doi.org/10.1080/15568318.2019.1573939
31. Saritas O, Meissner D, Sokolov A (2018) A transition management roadmap for fuel cell electric vehicles (FCEVs). J Knowl Econ. https://doi.org/10.1007/s13132-018-0523-3
32. Miles I, Saritas O, Sokolov A (2016) Foresight for science, technology and innovation. Springer International Publishing, Switzerland
33. Saritas O, Miles I (2012) Scan-4-light: a searchlight function horizon scanning and trend monitoring project. Foresight 14(6):489–510
34. Box GEP, Wilson KB (1951) On the experimental attainment of optimum conditions. J R Stat Soc 13(1):1–45

Part III
Making in India

This section is on the problems and challenges in making and assembling EVs in India. This part has three chapters. Chapter 4 is on the value chain of assembling and manufacturing in India. Chapter 5 is on the imports of raw materials like lithium and components like batteries which are not manufactured in the country. It also discusses the exports of components. Chapter 6 discusses investment flows in the industry and the problems in attracting FDI in the future.

Chapter 4
Value Chain for Manufacture and Assembly

4.1 Introduction by Editor

4.1.1 Introduction

The value chain of the EV for manufacture or assembly comprises several steps. Certain parts are common to both ICE and EV: e.g. clutch/torque control, start/stop system, pistons, spark plugs, and turbochargers. The new parts include a power distribution module, onboard charger, battery management system, vehicle interface control module, DC/DC converter, etc. The most critical part is the battery. Every battery has two electrodes —a cathode and an anode—between which trillions of charged lithium atoms travel (in case of the lithium-ion batteries). The cathode largely determines a battery's performance, cost, and environmental footprint. Cathode today is produced almost entirely in Asia. This chapter will look at India's status in assembly and future manufacturing of batteries.

4.1.2 Value Chain of an EV

The electric vehicle value chain has power entering the battery at one end while powering the wheels at the other end as shown in Fig. 4.1. The battery supplies power as the energy source to the motor which through the gearbox, differential, etc. transfers power to the wheel of the vehicle.

The critical elements of powertrain manufacturing are:

- Battery (typically the Li-Ion battery with technology selection LFP or NMC chemistry)
- Motors (type of motors to be selected as per application in the vehicle)
- Electronics and the intelligent system of a vehicle

S. Ray and D. Mukherji (eds.), *A Primer on Electric Vehicles in India*,
https://doi.org/10.1007/978-981-97-8966-5_4

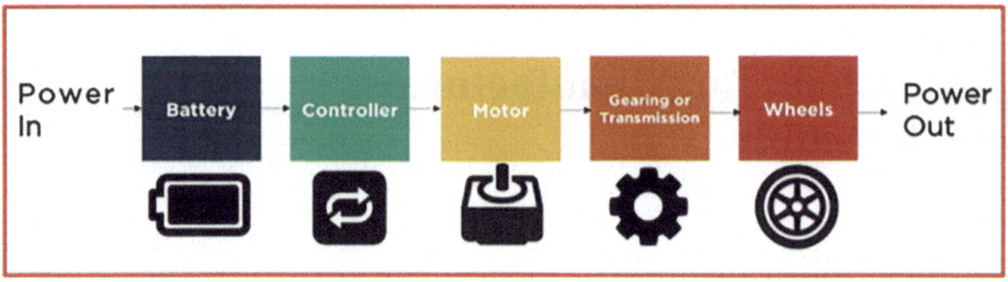

Fig. 4.1 Power train configuration for the EVs

The critical part of battery technology is managing the battery power efficiently. The battery management and control system (BMS) plays a major role in the efficient management of the EV powertrain. The BMS is an intelligent system that controls motor power as per torque requirement and also captures battery data e.g. energy consumption, charge/discharge cycles, thermal management of battery, etc.

In the electric vehicle, there are several new components (such as the power distribution module, DC-DC converter, vehicle interface control module, EV communication module, and battery management system (BMS), etc.) shown in the blue boxes in Fig. 4.2.

EV has only 20% of the parts compared to an ICE vehicle, so it is much simpler as well as more efficient technology.

The battery cell-making process involves several steps, and the manufacturing value chain is quite complex. The electrode production is the first step wherein it all starts with the mixture of various minerals in exact proportion, called slurry. The slurries consist of conductive additives, polymer binders, and electrochemically activated material particles that serve as reservoirs for Lithium. They are then coated onto conductive substates and dried to form porous electrodes. After electrodes are made, depending upon the type of cell—cylindrical, prismatic, or pouch type, current carriers of tabs are welded, insertion of the anode is done and finally the sealing and wrapping are done. The cell is now ready for testing, grading, and packaging for dispatch.

All these processes of cell manufacturing require specialized machinery and equipment. These technologies are currently with a handful of companies globally. Most of them are in China, Korea, Japan, and the USA. India is completely dependent on some of these companies for cell manufacturing technologies.

Process flow for Li-Ion battery manufacturing (Fig. 4.3).

4.1.3 Assembly and Manufacture in India

The automobile industry works at 3 levels with the vehicle maker OEM at the top. Below them are the Tier 1 suppliers who are essential system suppliers with complete R&D for product development and system integration. Below Tier 1 are levels 2 and

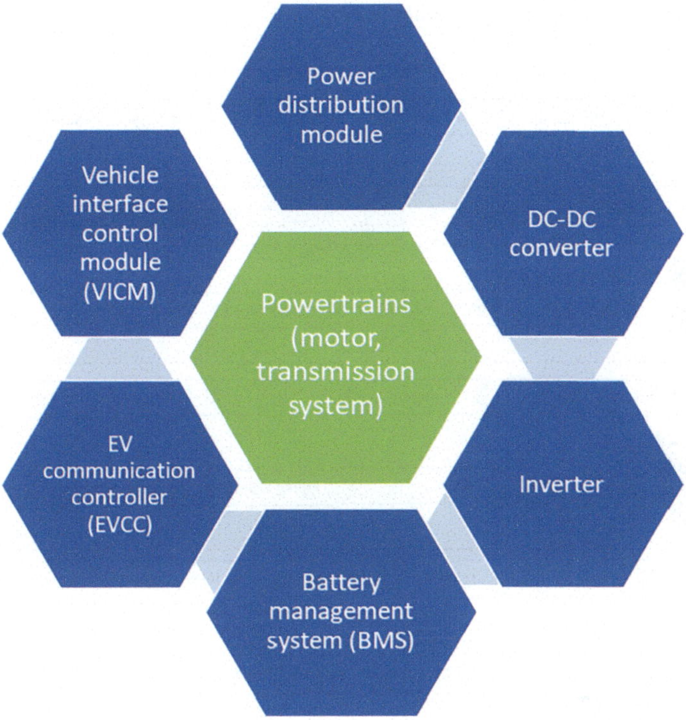

Fig. 4.2 Components of the EVs

3 which are specialized in their own domain parts. These companies are typically MSMEs.

Summary of India's Li-Ion cell making competency:

However, at the moment India has gaps in all of the components of EV manufacturing Battery and motor manufacturing. This is indicated in Table 4.1. As regards batteries, India has the handicap of not having most of the materials available in the country as has been discussed in Chap. 3. This supply chain needs to be set up by imports from other countries. Hence, cell or battery manufacturing is key to achieving large volumes of vehicle manufacturing.

It has been estimated that India will require 3400–4100 GWh of batteries by 2035 (Gode et al., 2021). Establishing a battery manufacturing plant has a long lead time as the supply chain is quite complex. As per McKinsey, an experienced battery maker requires 1 year to set up a Gigafactory while a new incumbent takes up to 4 years to set up the same Gigafactory. McKinsey reports, that only 70% of planned capacities will actually come into production so supply will reach 2 TWh, leading to a gap of 1.9 TWh. Securing the supply of key raw materials is key to the success of Gigafactory which is easier for experienced battery makers.

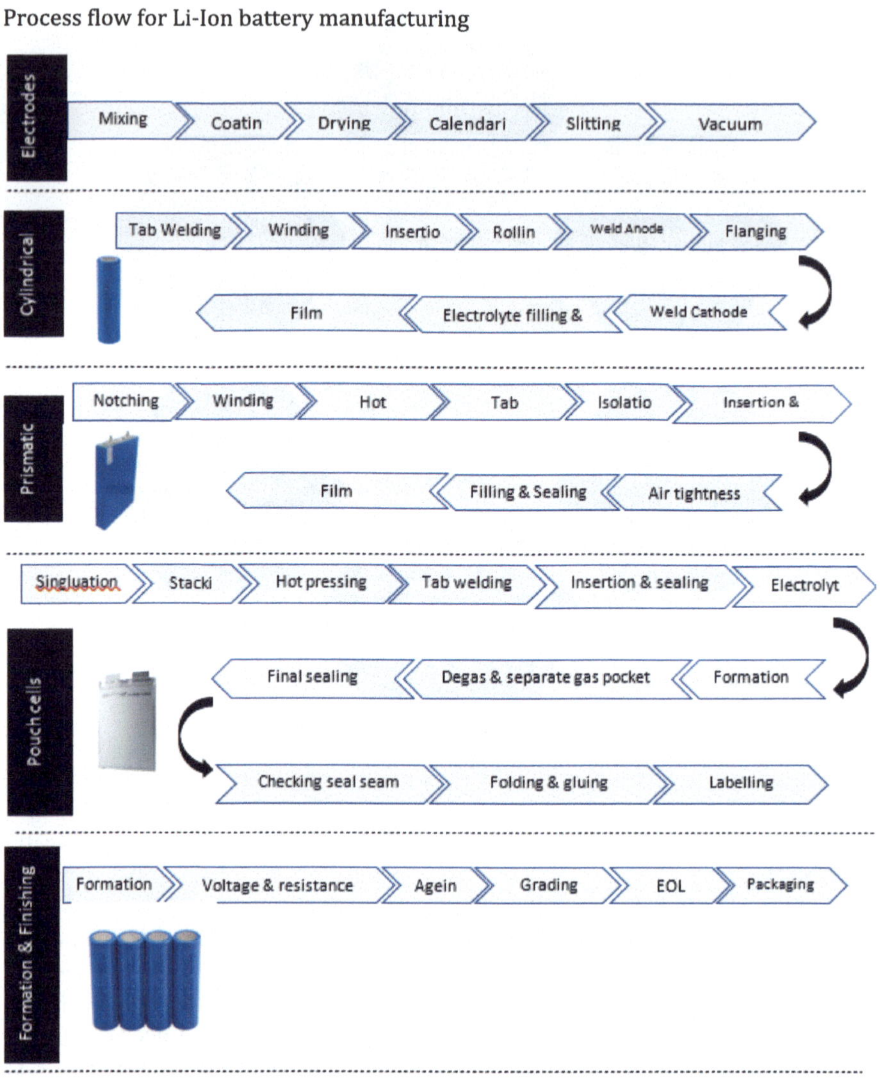

Fig. 4.3 Li-Ion cell manufacturing value chain (*Source* Authors' compilation)

4.1.4 Way Forward

As far as the other parts are concerned, India needs to set up an elaborate industrial base where we will have raw material suppliers at the base and then the suppliers of parts, and components for various industries higher in the value chain. Such a value chain would be self-sufficient [1] in materials, manpower, machinery, and finances.

Table 4.1 India's competency in Li-Ion battery manufacturing

	Domain	Details	Status	Remarks
1	Design and R&D	Cell design, Pack design for vehicles	Poor	Cell and battery pack level. BMS R&D is nonexistent
2	Materials	Lithium	NA	Major gaps in the supply chain
		Cobalt	NA	
		Manganese	NA	
		Graphite	Avail	
		Separators	NA	
		Polymers	NA	
		Aluminium	NA	Cell-grade metal is not available
		Copper	NA	Cell-grade metal is not available
3	Refining	Refining of Li, Co, Ni	NA	Refining facilities are not available
4	Cell manufacturing technology	Cell manufacturing	NA	Li-Ion cell manufacturing technology is not available in India
5	Battery pack manufacturing	Battery pack manufacturing with imported cells	Available	Battery packs are made in India with imported cells. These are assembled depending on the form factor required for vehicles
6	Testing and validation	Cell testing for grading, durability testing for life cycle analysis	NA	Cell-level testing facility is not available
		Thermal management testing	NA	Not available
	Infrastructure	Infrastructure for cell manufacturing	NA	Supply chain of crucial materials like Li, Co, and Mn is not available
		Infrastructure for pack manufacturing	2 GWH	Excide has recently started a Giga pack factory

Machine Generated

Ashok et al. (2022) examine the barriers to EV adoption and conclude that these barriers and costs will be reduced as battery technology evolves and increased density of charging infrastructure. Rajendran et al. (2020) explore the different types of energy-efficient converters for the assessment of power factor at electric vehicle charging stations. Abdul-Manan et al. (2022) estimate the life cycle emissions of 2-wheelers and 4-wheelers for the year 2018–19. Mathiyazhagan et al. (2017) analyze the criteria for choosing the green supplier base and find that manufacturing ranks last. Chaudhury and Bhasin (2012) predict a great future for EVs due to the emphasis on electric mobility.

References

Breiter, A., Horetsky, E., Linder, M., and R. Rettig (2022) EV Battery shortage: The market gets hotter. McKinsey.

Gode, P., Bieker, G., and Bandivadekar, A. (2021) Battery capacity needed to power electric vehicles in India from 2020 to 2035. International Council on Clean Transportation Working Paper no. 7.

Rajshekhar, M. (2023) Can India create an Atmanirbhar EV Manufacturing chain? Wire. July 30. Link: https://carboncopy.info/can-india-create-an-atmanirbhar-ev-manufacturing-chain/

Footnotes/Endnotes

[1] Rajshekhar (2023).

4.2 Machine Generated Summaries

Disclaimer: The summaries in this chapter were generated from Springer Nature publications using extractive AI auto-summarization: An extraction-based summarizer aims to identify the most important sentences of a text using an algorithm and uses those original sentences to create the auto-summary (unlike generative AI). As the constituted sentences are machine selected, they may not fully reflect the body of the work, so we strongly advise that the original content is read and cited. The auto generated summaries were curated by the editor to meet Springer Nature publication standards. To cite this content, please refer to the original papers.

Machine generated keywords: criterion, ghg emission, ghg, passenger, charger, electrification, supplier, selection, converter, evaluation, mitigation, shift, ecological, supply chain, life cycle.

4.2.1 Transition to Electric Mobility in India: Barriers Exploration and Pathways to Powertrain Shift Through MCDM Approach [1]

This is a machine-generated summary of:

Ashok, B.; Kannan, C.; Usman, Kaisan Muhammad; Vignesh, R.; Deepak, Chirag; Ramesh, Rathan; Narendhra, Tharun M. V.; Kavitha, C.: Transition to Electric Mobility in India: Barriers Exploration and Pathways to Powertrain Shift through MCDM Approach [1].

Published in: Journal of The Institution of Engineers (India): Series C (2022).

Link to original: https://doi.org/10.1007/s40032-022-00852-6.

If you want to cite the papers, please refer to the original.

For technical reasons we could not place the page where the original quote is coming from.

Abstract-Summary

"The ever-increasing environmental concern necessitates the implementation of electric vehicles (EVs) in the public fleet in a larger proportion."

"To support their future endeavours in the field of electric vehicles, a review article that encompasses the present status, barriers for EVs adoption and operation in developing countries proves highly critical."

"A critical analysis of EV adoption barrier information is used to develop appropriate technology to tackle these barriers and provide necessary plans for policymakers to effectively implement electric vehicles."

"For developing countries such as India, the model placed battery-operated electric vehicles at 5th position on the hierarchy ranking."

"The article also explores the prospects and critically analyses hybrid electric vehicles as an intermediary propulsion system that smoothens the transition to EV."

"The outcomes of the literature review, expert opinions and author findings suggest that hybrid technology uses powertrain components that are chiefly similar to those found in conventional vehicles, yielding vehicle manufacturers with a smoother changeover to Electric Vehicle technology, which may otherwise lead to an immense failure in technology transition."

Introduction

"Even a small disturbance or irregular transition from ICE-powered vehicles to EVs will produce a negative impact on the economic growth and overall performance of the automotive sector."

"These studies have often presented and compared comprehensive quantitative information for the emissions/fuel efficiency standards of conventional vehicles (CV), EVs and HEVs."

"A majority of research work highlights the importance of implementation of EVs and HEVs over CVs although most of them refrain from going into detail on comparing the two based on factors ranging from performance to infrastructure."

"Research work falling under the above categories neither consider the potential of other propulsion systems nor compare the overall characteristics of EVs to fossil liquid fuel vehicle, gaseous fuel vehicle, hybrid vehicle, plug-in hybrid vehicle, fuel-cell vehicle and alternative fuel vehicles."

Motivation and Objective

"This review article is focused on the motivation of elucidating the status of EVs with a detailed examination of the significant barriers to their adoption and operation in the context of developing countries."

"Included with the review status for an objective of identifying the possible alternative propulsion for the passenger vehicle segment, a quantitative examination method is proposed in the successive sections of the paper."

"To the best of the author's knowledge, there are no review papers that deal with a plan for an efficient propulsion system transition pathway."

"The existence of enormous barriers for imminent EV transition necessitates a decisive solution with the perspective of an intermediate propulsion system that is capable of utilizing the existing industrial and infrastructural facilities."

"Objective, an expert perception ranking method is utilized in discerning and justifying the mindset of respective experts on EVs and other propulsion types."

Electric Vehicle Transportation Status Around the Globe

"Developing nations such as India and Brazil have shown significant growth in the EV sector although their sales volumes are incomparable with the top three entities within this market [2]."

"In China, there exist adequate public charging points and hence the demand for EVs is less subservient to government policies as consumers face fewer issues with range anxiety and EV recharge."

"The gradual increase in projected sales of CVs till 2025 is ascribed to vehicle manufacturers' intention of improving their existent powertrains before the transition to EVs is completed."

"A target sale of 30–40 million EV units by 2030 is set by the Indian government, one-third of which is expected to be EV cars."

"Despite various initiatives taken by the Indian government, certain barriers limit the production and sales of EVs in India such as charging infrastructure, battery technology, consumer behaviour and overall government support which are elaborated in detail in the subsequent section."

Practical Challenges in Implementing EVs for Mass Transport

"Charging infrastructure, driving range, charging time, government support, total cost and awareness among consumers are crucial factors limiting EV adoption rates [3]."

"The present battery technology is a significant barrier that affects the overall cost, driving range and charging time of EVs [4]."

"This latter technology presents additional challenges such as increased installation and repair costs, inconsistent charging due to its diurnal nature and reduced battery life in the case of standalone photovoltaic cell based EVs. [5]."

"The effectiveness of EV batteries is determined by their range, cost, charging time and safety, working temperature range, materials used, recycling prospectus and disposal capabilities."

"The charging time of these batteries is much higher than the refuelling time of CVs, even considering fast charging technology [6], a significant barrier to EV adoption."

"Range, charging issues and high capital cost of EVs substantially influence consumer behaviour [7–9]."

Experts Perception towards Pathway of Vehicle Power Train Shift through MCDM Tool

"In such aspects, the main criteria: Economic, Behavioural, Infrastructure, Society and Environment are identified as the dominant decision-making variables in vehicle propulsion selection."

"By inducing changes in economic criteria weights while keeping all other criteria weights identical, liquid fuel vehicles (LFVs) are spotted as the best propulsion system for the present scenario of this economic phase."

"LFVs move to the lower zone as the entire weight is provided to the behavioural criteria, while BEVs top this position because of higher energy conversion efficiency levels."

"PHEV, BEV, AFV, and FCEV vehicles are ranked in subsequent positions as a significant change in charging infrastructure is required to make them functional."

"The PROMETHEE analysis is further proceeded as per the first phase of decision, and equivalent weights are employed for economic, behavioural, infrastructure, society, and environment-based criteria."

Hybrid Electric Vehicle Status

"There has been a considerable increase in the sales of HEV and PHEVs since the last decade [10]."

"This was followed by the USA with cumulative sales of 5.4 million units as of December 2019 [11], closely followed by Europe with around 550,000 units being sold in 2019 alone and cumulative sales of 3 million units of HEVs as of July 2020 [12]."

"A steady increase in the market share of PHEVs can be observed in France, Korea and the USA, with Europe displaying a similar trend."

"Sales of HEVs in the US began to decline following the financial crisis of 2007–08, and after a short recovery, began to decline again in 2014 due to low gasoline prices, when CVs dominated the market and later had a rebound in 2019 to 2.4% of the market share, along with PHEVs jumping to 0.5% of the market share [13]."

"This point is further supported by the fact that the sales of more than 15 million HEVs worldwide have resulted in savings of more than 120 million tonnes of CO_2 [12]."

Advantages of Hybridizing the existing ICE Infrastructure

"HEVs are more eco-friendly than CVs and EVs at least until a significant development in battery production and technology or electricity generation occurs through RES [14–16]."

"HEVs have lower TCO than EVs due to lower manufacturing costs, improved fuel efficiency, and government incentives [7]."

"EVs also get incentives from governments, but their TCO is higher than HEVs mainly because EVs have a high manufacturing cost, especially for their large batteries [17]."

"The upstream emission costs for energy production are very high for EVs as electricity generation is mainly from fossil fuels, further increasing its TCO [18–20]."

"The HEV engine also operates in its optimal condition, thereby improving its fuel economy and hence having a higher range than its CV counterparts and EVs."

"The bio fuelled HEVs follow a trend of lower brake power and torque, and higher brake specific consumption when compared to fossil fuel HEVs which necessitates the tuning of the engine operating parameters, thereby increasing cost [21, 22]."

Concluding Remarks

"The majority of the technical barriers and purchase costs could be reduced with corresponding development in battery technology and increased density of charging infrastructure."

"Governments of developing and underdeveloped countries should consider these barriers during the creation and implementation of new policies to maximize the EV transition."

"An expert perception survey conducted at the end of the review revealed the potential of HEVs as the best alternative propulsion system for India."

"The ample use of HEVs in India will help in reducing emissions and path transition to EVs."

"Literature review on the advantages of HEVs portrays its capability to ease EV transition."

"With increased HEV sales, production technology for electric-drive components increases and the supply chain for these components undergoes significant development."

"It helps to develop the charging infrastructure gradually and gives the consumers an experience of all-electric drive, thus helping EV adoption rates."

4.2.2 Energy-Efficient Converters for Electric Vehicle Charging Stations [23]

This is a machine-generated summary of:

Rajendran, Gowthamraj; Vaithilingam, Chockalingam Aravind; Naidu, Kanendra; Oruganti, Kameswara Satya Prakash: Energy-efficient converters for electric vehicle charging stations [23].

Published in: SN Applied Sciences (2020).

Link to original: https://doi.org/10.1007/s42452-020-2369-0.

If you want to cite the papers, please refer to the original.

For technical reasons we could not place the page where the original quote is coming from.

Abstract-Summary

"This article reviews the design and evaluation of different AC/DC converter topologies of the present status and future implementation plans for DC fast-charging infrastructures."

"This paper also evaluates the architecture, merit, and demerits of AC/DC converter topologies for DC fast-charging stations."

"Analysis, it has found that the Vienna rectifier is the best suitable converter topology for the high-power DC fast-charging infrastructure (> 20 kW), thanks to its low current ripples, low output voltage ripples, high efficiency, high power density, and high reliability."

"The paper focuses specifically on different topologies of Vienna rectifier topologies on Level-3 DC fast-charging stations which direct to less CO_2 emissions in electric vehicle charging stations, thus contributing to sustainable development goals of climatic action."

Introduction

"Level 3 chargers are typically designed for a fast charging using DC with the power rating around 100 kW with a charging time of less than 30 min."

"DC fast-charging stations for EVs are designed to supply about 50 kW of power [24]."

"Unidirectional chargers can be controlled to charge the EV battery from the grid [25–27]."

"As per the previous review papers [28–33], they have reviewed two-level AC–DC converters, conventional boost rectifier, zero-voltage transition (ZVT) converters, zero-current transition (ZCT) converters, ZVT-ZCT converters, interleaved boost PFC converters, bridgeless boost PFC converters, and bridgeless interleaved boost PFC converters for the EV charging stations based on the efficiency, power factor, and input current THD and this paper reviews the practical viability of the energy-efficient converters based on the efficiency, power factor, power density, input current THD, and simulation analysis of Vienna rectifier for EV charging stations is carried out."

DC Fast-Charging Converter Topologies

"The primary goal of using a boost converter instead of the conventional diode bridge rectifier is to improve power factor, to reduce the harmonics at the end, and to have a controlled DC voltage at the output if unwanted perturbations occur at the AC end."

"Compared to boost-type converter, buck-type system provides a wide output voltage control range, while maintaining PFC capability in the input, enables direct start-up, and allows for dynamic current limitations at the output."

"Matrix converter is a forced commutated converter that uses an array of controlled bidirectional switches which allows high-frequency operations."

"This converter consists of only one active switch per phase which makes the Vienna rectifier easier to control and makes it more dependable."

"This converter is basically a PWM converter [34], and the boost inductor at the input plays the role in ascertaining power factor correction."

Performance Comparison of Energy-Efficient Converters

"Good efficiency when compared to the PWM rectifier, SWISS rectifier, and matrix converter."

"Higher power factor, around 0.99, compared to the PWM rectifier, SWISS rectifier, and matrix converter."

"The Vienna rectifier is the most optimal converter topology for the charging stations as it has the advantages of high power density (12 kW/dm^3) [35, 36] compared to the SWISS rectifier (4 kW/dm^3) [37]."

Control Strategies of Energy-Efficient Converters

"Several studies have established that model predictive control (MPC) reduces the harmonics in the line current and a smaller mean absolute current reference tracking error as compared to other controllers."

"In [38], the author presented a predictive current control method for reducing the total harmonic distortion (THD) by using a switching frequency of 8 kHz with a voltage source inverter."

"In [39], the researcher applied the model predictive control algorithm for a four-leg converter to observe the reduction in THD and switching frequency at low values of the filter parameters."

"In the single-phase on-board bidirectional charger proposed by [25], PI controllers were employed in AC/DC converters and DC/DC converters which provide constant voltage and constant current charging, as well as reactive power compensation."

"This PFC controller has two outer voltage loops and one inner current loop."

Topologies of Vienna Rectifier for Charging Stations

"The key characteristics of this converter known as split capacitor consists of two capacitors being placed at the output side, that reduces the voltage stress on to the power semiconductor switches."

"This topology has only one semiconducting switch and six diodes which reduce the efficiency of the system."

"To reduce the voltage stress on the switches, two capacitors are connected in parallel to minimize the losses in the switches [40]."

"In Topology 2, the voltage ripple is minimized by balancing the voltage using two switches connected in anti-parallel to the neutral."

"As the switch is made up of two MOSFETs, it is different from the Topology 1 which reduces the voltage stress on the switches and it has only two diodes in the circuit which reduces the losses in the diodes."

Results and Analysis

"In all three topologies, there are slight changes in the harmonics, which leads to minor variations in the sinusoidal waveform of input current."

"It is noted that the percentage input current THD for Topology 1 is 4.11, which is less than 5%, and it is meeting the IEEE standards."

"Due to the more number of switches in Topology 2, the harmonics on the input current are less compared to Topology 1 which is 6.45% and it fails to meet the IEEE standards."

"The percentage input current THD for Topology 3 is 2.38, which is less than 5%, and it satisfies the IEEE standards."

"Due to this, the harmonics on the input current in Topology 2 increased to more than 5% which affects the power quality of the source."

Conclusion

"This article comprehensively explore different types of energy efficient converter topologies for power factor assessment at the electric vehicle charging stations."

"Three topologies of the Vienna rectifier were compared using the following parameters.1.Number of active/passive devices.2.Total loss and efficiency.3.Input current THD.4.Power factor."

"The performance of three topologies is simulated and analysed in terms of a number of active/passive devices, total loss and efficiency, power factor, and THD."

"It can be concluded that the Topology 3 is the most suitable converter for the electric vehicle charging stations and it is benchmarked for less complexity, high efficiency, high power density design, less input current THD, and improved power factor."

4.2.3 Comparative Life Cycle GHG Emission Analysis of Conventional and Electric Vehicles in India [41]

This is a machine-generated summary of:

Das, Jani: Comparative life cycle GHG emission analysis of conventional and electric vehicles in India [41].

Published in: Environment, Development and Sustainability (2022).

Link to original: https://doi.org/10.1007/s10668-021-01990-0.

If you want to cite the papers, please refer to the original.

For technical reasons we could not place the page where the original quote is coming from.

Abstract-Summary

"The life cycle GHG emissions of electric vehicles in terms of equivalent carbon emission ($kgCO_{2eq}$) are compared with conventional vehicles for a life cycle inventory in Indian conditions."

"It has been concluded that there is a reduction of about 40% embodied equivalent carbon in an ICEV in comparison with an EV in Indian conditions."

"Vehicle emission factor has been introduced to normalize the emission value with respect to the vehicle life in km."

"The study concludes that the high value of emission parameters for EVs in comparison with ICEVs is due to the high rated battery component, emission factor due to the Indian generation mix and energy intensive manufacturing techniques."

"The emission footprint of EVs can be reduced by improved material production and manufacturing techniques of the vehicle components, increased penetration of renewable in the generation sector and enhanced usage of recycled components in vehicle industry."

Introduction

"The report also states that the usage of internal combustion engine vehicles will decline globally, peaking in the late 2020s, with around 330 million electric cars on roads by 2040."

"According to the report by Innovation Norway (India EV Story, 39) the key drivers for the advent of Electric Vehicles in India are high on road carbon emissions, lower power demand and fuel security risks."

"This study contributes to the existing LCA studies already published in the literature in the following: life cycle GHG emission analysis of transportation technology involving a Cradle to Grave (CTG) and End of Life (EOL) approach in India; compilation of Life Cycle Inventory (LCI) purely in Indian conditions, making the study relevant to the location; extension of analysis for a projected location specific electricity generation scenario; sensitivity studies to battery manufacturing techniques and electricity generation mix."

Literature Review

"The significance of lithium ion batteries in EV life cycle energy and emissions and their recycling has been studied and presented by Dunn and others (40)."

"Yang and others (41) has studied the impact of battery degradation on life cycle GHG emission of electric vehicles."

"Main limitations identified in the available literature are: lack of estimated LCI specific to a region calculated from basic material and energy flows; lack of a complete life cycle energy and environmental analysis in terms of GHG emissions of any vehicle technology in India with Indian data; lack of studies related to impact of electricity generation mixes in the present and future scenario on life cycle GHG emissions of a vehicle technology in India."

"This research studies the impact of future generation mix of India for two scenarios, on life cycle GHG emissions of vehicle technologies."

Methods

"The specific aim and scope of this study can be summarised as: to provide LCI data for two fuel technologies of an ICEV and three battery variants of an EV in Indian conditions; to estimate component-wise, life cycle phase-wise and material-wise GHG and CO_2 emission values of the vehicles from the estimated LCI data; to study the impact of present and future electricity generation scenarios for India; to present comparative analysis between the different vehicle types; to present sensitivity analysis to generation mix and battery manufacturing techniques; to suggest mitigation opportunities."

"The goal of this study is to evaluate and conduct a comparative GHG emission analysis associated with material production, manufacturing, use and disposal of conventional and electric vehicles in an Indian scenario."

"The approach is aimed at identifying key factors related to the energy and GHG emissions in different life cycle stages of vehicles with selected traction batteries."

Results and Discussion

"Steel and its associated materials contribute about 21% for ICEV, 17% for LFP variant, 27% for NMC variant and 33% for LMO variant of the total material GHG emissions."

"For the EV variants, the contribution of battery active materials and graphite, the anode material also contribute in a varied range to the total GHG emissions."

"The projected value for an optimistic scenario of 27% renewable share in electricity generation the year 2026–2027 in the country from present 21% (including large hydro) is depicted by a GHG emission factor of 0.9 $kgCO_{2eq}/kWh$ according to the study conducted by The Energy and Resources Institute (Transitions in Indian Electricity Sector), which gets accounted for, using the parameter 0.75 in the GHG footprint of the present scenario for the reference scenario and 0.65 for the REmap Scenario."

Conclusion and Future Scope

"Air pollutants are less during the driving phase of an EV, the vehicle life cycle emissions are higher than that of ICEVs in the present scenario."

"The GHG vehicle emission factor for ICEV petrol variant is 0.27 $kgCO_{2eq}$/km and 0.24 $kgCO_{2eq}$/km for the diesel variant."

"The Cradle to Gate GHG emission factor of petrol and diesel variants are the same, 0.0436 $kgCO_{2eq}$/km assuming similar materials and processes for manufacturing the vehicle."

"If the raw materials are imported and improved and less energy intensive manufacturing methods are adopted inhouse, the battery contribution to the emission footprint of the cradle to gate of the vehicle can be reduced."

"The ambiguity relating to the manufacturing techniques of vehicle components especially the lithium ion battery, the vehicle specifications like mileage and vehicle life and the GHG footprint of electricity generation, will have impact on the results presented."

4.2.4 Electrifying Passenger Road Transport in India Requires Near-Term Electricity Grid Decarbonisation [42]

This is a machine-generated summary of:

Abdul-Manan, Amir F. N.; Gordillo Zavaleta, Victor; Agarwal, Avinash Kumar; Kalghatgi, Gautam; Amer, Amer A.: Electrifying passenger road transport in India requires near-term electricity grid decarbonisation [42].

Published in: Nature Communications (2022).

Link to original: https://doi.org/10.1038/s41467-022-29620-x.

If you want to cite the papers, please refer to the original.

For technical reasons we could not place the page where the original quote is coming from.

Abstract-Summary

"Battery-electric vehicles (BEV) have emerged as a favoured technology solution to mitigate transport greenhouse gas (GHG) emissions in many non-Annex 1 countries, including India."

"GHG mitigation potentials of electric 4-wheelers in India depend critically on when and where they are charged: 40% reduction in the north-eastern states and more than 15% increase in the eastern/western regions today, with higher overall GHGs emitted when charged overnight and in the summer."

"Self-charging gasoline-electric hybrids can lead to 33% GHG reductions, though they haven't been fully considered a mitigation option in India."

Introduction

"As one of the participating countries, India recently launched phase 2 of the Faster Adoption and Manufacturing of Electric Vehicles (FAME II) scheme, which provides 1.3 billion USD fiscal incentives over the next three years to support the rapid electrification of the transport sector [43]."

"After the power generation and industrial sectors, India's transport sector is the third-largest energy-related GHG emission source accounting for about 13% in 2019 [44]."

"We model the life cycle GHG emissions of over 600 passenger road vehicles comprising two-wheelers (2 W) and four-wheelers (4 W) available in India in the financial year 2018/19."

"We assess the life cycle GHG emission ranges of passenger road vehicles in India by powertrain types."

"We conclude that electrification can offer climate change mitigation potential in some sectors and regions, but it is not the panacea for India's passenger road transport."

Results

"A more rapid reduction in the electric grid carbon intensity could enable a larger reduction in the life cycle GHG emissions of the BEV."

"Phase 2 of the fuel consumption standard in India, which comes into effect from the year 2022/2023 onwards [45], stipulates that 4 W passenger vehicles with an average curb weight of 1237 kg should not exceed 4.95 L/100 km."

"A BEV less efficient would have higher overall GHG emissions than a comparable gasoline 4 W vehicle."

"Strategies to electrify future passenger transport in India must involve developing energy-efficient BEVs and deep decarbonisation of the power sector to ensure an overall reduction in GHG emissions."

"The three most important assumptions/simplifications in this study include: (1) the adoption of a static electricity mix for India, (2) the uncertain real-world driving impacts, particularly for a highly-electrified vehicle, and (3) the absence of a vehicle fleet dynamics."

Discussion

"An organised mobility transition plan supported by phased electrification focused on 2-wheelers, representing 80% of India's passenger road vehicles, is needed."

"This highlights the importance of understanding the trade-offs in policy decisions and the need for a systemic approach to guide India's overall sustainable transition in the transport sector."

"A sustainable transport policy has to incentivise the right technology mix at the right time and in the right location."

"Effective behaviour-centric policies are needed to attenuate demand for road vehicle travel [46] through (i) support for active mobility (cycling and walking), (ii) enhancement of public transit, (iii) optimal design of the built environment, and (iv) reducing the need for travel (telecommute and telework)."

"We need a mosaic of technologies for different transport sectors and geographical regions, including electrified powertrains (counting full BEV), highly efficient combustion engine-powered vehicles, and low-carbon fuels such as biofuels, hydrogen, and CO_2-derived fuels, including Methanol and DME."

Methods

"We derived this using the official fuel mix and power generation data from India's Central Electric Authority (CEA), an organisation under the Ministry of Power [47]."

"It is estimated to supply only 31% of India's overall electricity generation mix in 2029/30 [48]."

"This study utilised vehicle energy consumptions and exhaust emissions data, certified by the Automotive Research Association of India (ARAI)."

"Type approvals for vehicle energy consumption and resulting tailpipe emissions, particularly under the NEDC drive cycle, are known to underestimate the real-world efficiency and emissions by an average margin of 10–40% based on empirical findings in the USA, Europe, and China [49–54]."

"For a class-1 2-wheeler with a curb weight of 109 kg, the net production and recycling emissions were estimated to be about 1.1 ton CO_{2eq}, resulting in an overall life cycle GHG emissions of about 73.8 gCO_{2eq}/km."

4.2.5 Modeling the Criteria for Selection of Suppliers Towards Green Aspect: A Case in Indian Automobile Industry [55]

This is a machine-generated summary of:

Mathiyazhagan, K.; Sudhakar, S.; Bhalotia, Abhishek: Modeling the criteria for selection of suppliers towards green aspect: a case in Indian automobile industry [55].

Published in: OPSEARCH (2017).

Link to original: https://doi.org/10.1007/s12597-017-0315-8.

If you want to cite the papers, please refer to the original.

For technical reasons we could not place the page where the original quote is coming from.

Abstract-Summary

"Environmental practices are not only based on the industry alone, supplier is also one of the partners to improve industries environmental image through supplying environmental friendly products."

"Selection of a suitable supplier based on environmental criteria is a challenging and difficult task for industries because there are many criteria available for selection of suitable environment friendly suppliers to improve industry environmental image."

"This study investigates and ranks the environmental criteria for the green supplier selection based on expert's judgment in an Indian automobile sector perspective."

Introduction

"Industrial environmental improvements depend upon the selection of suitable supplier based GSCM criteria aspects so that the firm's environmental sustainability and ecological performance can be demonstrated by its suppliers [56]."

"Selection of supplier in green aspect is a testing assignment for every one of the ventures who need to enhance their green image by producing environmental friendly products."

"Many different environmental criteria are available for selection of supplier because different industries and different countries have different opinions for same criteria based on their culture, political background, regulations, etc. [57, 58]."

"The objective of the paper is framed to analyze and rank the environmental criteria for green supplier selection from the substantial literature and rank the predominant environmental criteria from the recommended listing."

Literature Review

"Since a decade ago, businesses are more engaged towards executing the novel and viable methodologies for decreasing the pollution, effluences and related harmful entities in supply chain management (SCM)."

"Numerous researchers and practitioners developed the potential and efficient initiative for environmental practices in supply chain management [59–62], so that the ecological challenges, including waste disposal, air pollution, water pollution, acid rains, global warming and related risks can be handled and avoided with higher productivity."

"This work produces the general execution scope related with each supplier with the integration of Linguistic Preferences using AHP."

"An effective mathematical model is used to depict the efficiency of the proposed work."

Research Gap

"Supplier selection criteria differ according to country's environmental regulations."

"All the researchers have analyzed the criteria for green supplier selection with the help of different approaches."

"There is no research for ranking the important green supplier criteria and also there is less research in Indian context to analyze the criteria for the green supplier selection."

"Only few studies have analyzed the criteria for green supplier selection [63, 64]."

"These reviews are not engaged to name the vital green supplier criteria in broad view."

Illustration of Case

"Because of rivalry in the market, each industry has eminent awareness about the most recent innovation of their field to enhance their improvements through fulfilling their clients and keeping up a great association with them."

"India has a tremendous industry, delivering items and administrations to clients in light of their prerequisites."

"The supplier has a vital role to improve economically as well as environmentally by supplying the products in the perspective of environmental aspects without violating the economy [65]."

"Determination of supplier is a testing assignment for industry to get a pollution free items without violating the financial emergencies."

"Manager of XYZ Company wants to select the above mentioned 15 industries as suppliers for their company based on the important criteria by means of ranking by using the AHP approach."

Solution Methodology

"AHP is a multi-criteria decision making (MCDM) approach to rank the factors based on expert judgement."

"The collected data is processed, synthesized and finally analyzed using AHP technique."

"Target, Explanatory Analytic Hierarchy Process (AHP) approach is reasonable to rank the criteria."

"AHP is a well-established multi criteria decision making (MCDM) technique and broadly connected to enterprises to take the choice."

"Saaty [66] had introduced AHP to prioritize the criteria or alternatives based on criterion in a natural, pairwise comparison matrix."

"The underlying stride takes the identification of environmental criteria for structuring a model for hierarchy prioritization and step 2 follows the construction of the appropriate questionnaire and data collection in automobile industry perspective."

Questionnaire Design and Experts Opinions Collection

"Poll for this review was figured to gather the information from 35 recognized experts in 15 vehicle ventures."

"Be that as it may, just 15 ventures shared enthusiasm for this review after rehashed suggestions to industry delegates and sends with respect to the information gathering."

"Each industry recommended more than 2 experts based on their expertise related to the research objectives of environmental management, purchasing and the quality departments."

"From the experts, information was gathered in view of pairwise examination between criteria class and specific criteria."

Results and Discussions

"Bouzon and others [67] have pointed out that environmental initiation does not focus only on manufacturers, but also entire supply chain management manufacturers ought to understand the significance of enhancing their condition by actualizing management practices that reuse basic assets.. It is clearly mentioned that all the departments which are involved in supply chain activities are responsible to adopt the green (environmental) concept in their traditional activities."

"This paper names the criteria categories and specific criteria for the selection of green supplier based on AHP approach."

"Based on more recent developments in industries, technology category is acting as a primary role for the selection of a supplier."

"Five criteria are listed under management, namely Ecological Protection Policy and Objective (M1), Green safety certification (M2), Education and training (M3), Environmental fortification associate (M4) and Executive support (M5 Among the five criteria, M1 is going about as a key measure."

"Procurement of environmental friendly objects (MF4) criterion receives a third rank in this category."

Conclusions and Remarks

"Manufacturing category ranks last revealing that industry, though involved in adoption of greener products in their manufacturing processes and eliminating waste into the environment, yet at the same time confront a little gap which finally demonstrates that India is still in the underlying phase of embracing ecologically agreeable choice criteria in supplier choice."

"This research has given most essential green criteria requiring for selection of suppliers in the context of the Indian automotive industry."

"The paper has given Indian automotive industry's most important green criteria to select its suppliers and procure its material in the most ecological way."

"Only 15 criteria under four categories for supplier's selection towards green aspect were considered with the help of literature and experts discussion."

"The result of this research helps in adopting ecological prospect while selecting a supplier in context to India automotive scenario."

Bibliography

1. Ashok B, Kannan C, Usman KM, Vignesh R, Deepak C, Ramesh R, Narendhra TMV, Kavitha C (2022) Transition to electric mobility in India: barriers exploration and pathways to powertrain shift through MCDM approach. J Inst Eng (India): Ser C. https://doi.org/10.1007/s40032-022-00852-6
2. Canitez F, Alpkokin P, Kiremitci ST (2020) Sustainable urban mobility in Istanbul: challenges and prospects. Case Stud Transp Policy 8(4):1148–1157
3. He X, Zhan W, Hu Y (2018) Consumer purchase intention of electric vehicles in China: the roles of perception and personality. J Clean Prod 204:1060–1069. https://doi.org/10.1016/j.jclepro.2018.08.260

4. Hannan MA, Azidin FA, Mohamed A (2014) Hybrid electric vehicles and their challenges: a review. Renew Sustain Energy Rev 29:135–150. https://doi.org/10.1016/j.rser.2013.08.097
5. Ghosh A (2020) Possibilities and challenges for the inclusion of the Electric Vehicle (EV) to reduce the carbon footprint in the transport sector: a review. Energies 13(10):2602. https://doi.org/10.3390/en13102602
6. Shareef H, Islam MM, Mohamed A (2016) A review of the stage-of-the-art charging technologies, placement methodologies, and impacts of electric vehicles. Renew Sustain Energy Rev 64:403–420. https://doi.org/10.1016/j.rser.2016.06.033
7. Bjerkan KY, Nørbech TE, Nordtømme ME (2016) Incentives for promoting battery electric vehicle (BEV) adoption in Norway. Transp Res Part D: Transp Environ 43:169–180. https://doi.org/10.1016/j.trd.2015.12.002
8. Junquera B, Moreno B, Álvarez R (2016) Analyzing consumer attitudes towards electric vehicle purchasing intentions in Spain: technological limitations and vehicle confidence. Technol Forecast Soc Chang 109:6–14. https://doi.org/10.1016/j.techfore.2016.05.006
9. Lebeau K, Van Mierlo J, Lebeau P, Mairesse O, Macharis C (2013) Consumer attitudes towards battery electric vehicles: a large-scale survey. Int J Electr Hybrid Veh 5(1):28–41. https://doi.org/10.1504/ijehv.2013.053466
10. Evolution of market share of hybrid electric vehicles and plug-in electric cars in California (2011 through 2020)—California New Car Dealer Association (CNCDA)
11. http://www.eagleaid.com/AID-Newsletter-preorder-1703preview-f-Japan-Global-car-hybrid-capital.htm . Accessed 19 Mar 2021
12. https://tedb.ornl.gov/wp-content/uploads/2021/02/TEDB_Ed_39.pdf#page=57. Accessed 19 Mar 2021
13. https://www.autocarpro.in/news-international/toyota-drives-past-three-million-hybrid-ev-sales-in-europe-56849. Accessed 19 Mar 2021
14. Khalfan A, Andrews G, Li H (2017) Real world driving: emissions in highly congested traffic. SAE Tech Pap Ser. https://doi.org/10.4271/2017-01-2388
15. Tamayao MAM, Michalek JJ, Hendrickson C, Azevedo IM (2015) Regional variability and uncertainty of electric vehicle life cycle CO_2 emissions across the United States. Environ Sci Technol 49(14):8844–8855. https://doi.org/10.1021/acs.est.5b00815
16. Szumska E, Pawełczyk M, Pistek V (2019) Evaluation of the life cycle costs for urban buses equipped with conventional and hybrid drive trains. Archiwum Motoryzacji. https://doi.org/10.14669/AM.VOL83.ART5
17. Redelbach M, Propfe B, Friedrich HE (2012) Competitive cost analysis of alternative powertrain technologies. In: Conference Programme IAMF 2012
18. Xiong W, Wu Z, Yin C, Chen L (2008) Economical comparison of three hybrid electric car solutions. In: 2008 IEEE vehicle power and propulsion conference (IEEE, 2008), pp 1–6. https://doi.org/10.1109/VPPC.2008.4677402
19. Cicconi P, Germani M, Landi D, Mengarelli M (2014) Life cycle cost from consumer side: A comparison between traditional and ecological vehicles. In: 2014 IEEE international energy conference (ENERGYCON) (IEEE, 2014), pp 1440–1445. https://doi.org/10.1109/ENERGYCON.2014.6850612
20. Gilmore EA, Lave LB (2013) Comparing resale prices and total cost of ownership for gasoline, hybrid and diesel passenger cars and trucks. Transp Policy 27:200–208. https://doi.org/10.1016/j.tranpol.2012.12.007
21. Shirneshan AR, Almassi M, Ghobadian B, Najafi GH (2014) Investigating the effects of biodiesel from waste cooking oil and engine operating conditions on the diesel engine performance by response surface methodology. Iran J Sci Technol Trans Mech Eng 38(2):289
22. Shirneshan A, Samani BH, Ghobadian B (2016) Optimization of biodiesel percentage in fuel mixture and engine operating conditions for diesel engine performance and emission characteristics by Artificial Bees Colony Algorithm. Fuel 15(184):518–526
23. Rajendran G, Vaithilingam CA, Naidu K, Oruganti KSP (2020) Energy-efficient converters for electric vehicle charging stations. SN Appl Sci. https://doi.org/10.1007/s42452-020-2369-0

24. Wang S, Crosier R, Chu Y (2012) Investigating the power architectures and circuit topologies for megawatt superfast electric vehicle charging stations with enhanced grid support functionality. In: 2012 IEEE international electric vehicle conference, pp 1–8
25. Kisacikoglu MC, Kesler M, Tolbert LM (2015) Single-phase on-board bidirectional PEV charger for V2G reactive power operation. IEEE Trans Smart Grid 6:767–775
26. Monteiro V, Pinto JG, Afonso JL (2016) Operation modes for the electric vehicle in smart grids and smart homes: present and proposed modes. IEEE Trans Veh Technol 65:1007–1020
27. Tashakor N, Farjah E, Ghanbari T (2017) A bidirectional battery charger with modular integrated charge equalization circuit. IEEE Trans Power Electron 32:2133–2145
28. Tran VT, Sutanto D, Muttaqi KM (2017) The state of the art of battery charging infrastructure for electrical vehicles: topologies, power control strategies, and future trend. In: 2017 Australasian Universities Power Engineering Conference (AUPEC), Melbourne, VIC, 2017, pp 1–6. https://doi.org/10.1109/AUPEC.2017.8282421
29. Praneeth AVJS, Williamson SS (2018) A review of front end AC–DC. In: 2018 IEEE transportation electrification conference and expo (ITEC), Long Beach, CA, 2018, pp 293–298. https://doi.org/10.1109/ITEC.2018.8450186
30. Koushki B, Safaee A, Jain P, Bakhshai A (2014) Review and comparison of bi-directional AC–DC converters with V2G capability for onboard EV and HEV. In: 2014 IEEE transportation electrification conference and expo (ITEC), Dearborn, MI, 2014, pp 1–6. https://doi.org/10.1109/ITEC.2014.6861779
31. Na T, Yuan X, Tang J, Zhang Q (2019) A review of on-board integrated charger for electric vehicles and a new solution. In: 2019 IEEE 10th international symposium on power electronics for distributed generation systems (PEDG), Xi'an, China, pp 693–699. https://doi.org/10.1109/PEDG.2019.8807565
32. Kumar S, Usman A (2018) A review of converter topologies for battery charging applications in plug-in hybrid electric vehicles. In: 2018 IEEE industry applications society annual meeting (IAS), Portland, OR, pp 1–9. https://doi.org/10.1109/IAS.2018.8544609
33. Na T, Yuan X, Tang J, Zhang Q (2019) A review of onboard integrated electric vehicles charger and a new single-phase integrated charger. CPSS Trans Power Electron Appl 4(4):288–298. https://doi.org/10.24295/CPSSTPEA.2019.00027
34. Chelladurai J, Vinod B (2015) Performance evaluation of three phase scalar controlled PWM rectifier using different carrier and modulating signal. J Eng Sci Technol 10(4):420–433
35. Hang L, Zhang M, Tolbert LM, Lu Z (2013) Digitized feedforward compensation method Vienna PFC converter. IEEE Trans Ind Electron 60:1512–1519
36. Wang L, Zhang D, Wang Y, Gu Y (2015) Dynamic performance optimization for high-power density three-phase Vienna PFC rectifier. In: 2015 IEEE 2nd international future energy electronics conference, pp 1–4
37. Kedjar B, Kanaan HY, Al-Haddad K (2014) Vienna rectifier with power quality added function. IEEE Trans Ind Electron 61:3847–3856
38. Pontt J, Correa P, Rodríguez J, Member S, Pontt J, Member S, Silva CA (2007) Predictive current control of a voltage source predictive current control of a voltage source inverter. IEEE Trans Ind Electron 54:495–503
39. Rivera M, Yaramasu V, Rodriguez J, Wu B (2013) Model predictive current control of two-level four-leg inverters -Part ii: experimental implementation and validation. IEEE Trans Power Electron 28:1–9
40. Thandapani T, Karpagam R, Paramasivam S (2016) Comparative study of Vienna rectifier topologies. Int J Power Electron 7:147
41. Das J (2022) Comparative life cycle GHG emission analysis of conventional and electric vehicles in India. Environ Dev Sustain. https://doi.org/10.1007/s10668-021-01990-0
42. Abdul-Manan AFN, Gordillo Zavaleta V, Agarwal AK, Kalghatgi G, Amer AA (2022) Electrifying passenger road transport in India requires near-term electricity grid decarbonisation. Nat Commun. https://doi.org/10.1038/s41467-022-29620-x

43. GOI (2019) Scheme for faster adoption and manufacturing of electric vehicles in India Phase II (FAME India Phase II). Government of India, Ministry of Heavy Industries & Public Enterprises, Department of Heavy Industry, New Delhi. https://dhi.nic.in/writereaddata/UploadFile/publicationNotificationFAME%20II%208March2019.pdf
44. IEA (2021) India energy outlook 2021. International Energy Agency
45. ICCT (2020) Light-duty vehicle fuel consumption standards factsheet India. The International Council on Clean Transportation
46. Axsen J, Plotz P, Wolinetz M (2020) Crafting strong, integrated policy mixes for deep CO_2 mitigation in road transport. Nat Clim Change 10:809–818. https://doi.org/10.1038/s41558-020-0877-y
47. CEA, Data from the central electricity authority, ministry of power. Government of India [Online]. Available: https://cea.nic.in/?lang=en
48. CEA (2020) Report on optimal generation capacity mix for 2029-30. Central Electricity Authority (CEA), Ministry of Power, Government of India
49. Fontaras G, Dilara P (2012) The evolution of European passenger car characteristics 2000–2010 and ts effects on real-world CO_2 emissions and CO_2 reduction policy. Energy Policy 49:719–730
50. Pavlovic J, Marotta A, Ciuffo B (2016) CO_2 emissions and energy demands of vehicles tested under the NEDC and the new WLTP type approval test procedures. Appl Energy 177:661–670
51. Ntziachristos L et al (2014) In-use vs. type-approval fuel consumption of current passenger cars in Europe. Energy Policy 67:403–411
52. Kung L, Butler T, Georges G, Boulouchos K (2019) How much energy does a car need on the road? Appl Energy 256:113948
53. Fontaras G, Zacharof N-G, Ciuffo B (2017) Fuel consumption and CO_2 emissions from passenger cars in Europe—laboratory versus real-world emissions. Prog Energy Combust Sci 60:97–131
54. Weiss M, Cloos KC, Helmers E (2020) Energy efficiency trade-offs in small to large electric vehicles. Environ Sci Eur 32(46):1–17
55. Mathiyazhagan K, Sudhakar S, Bhalotia A (2017) Modeling the criteria for selection of suppliers towards green aspect: a case in Indian automobile industry. Opsearch. https://doi.org/10.1007/s12597-017-0315-8
56. Kuo RJ, Wang YC, Tien FC (2010) Integration of artificial neural network and MADA methods for green supplier selection. J Clean Prod 18(12):1161–1170. https://doi.org/10.1016/j.jclepro.2010.03.020
57. Mathiyazhagan K (2014) Govindan, Kannan, Haq, Noorul: pressure analysis for green supply chain management implementation in Indian industries using analytic hierarchy process. Int J Prod Res 52(1):188–202
58. Zhu Q, Sarkis J (2006) An inter-sectoral comparison of green supply chain management in China: drivers and practices. J Clean Prod 14(5):472–486
59. Ala-Harja H, Helo P (2014) Green supply chain decisions–Case-based performance analysis from the food industry. Transp Res E Logist Transp Rev 69:97–107
60. Fahimnia B, Sarkis J, Davarzani H (2015) Green supply chain management: a review and bibliometric analysis. Int J Prod Econ 162:101–114
61. Lee VH, Ooi KB, Chong AYL, Seow C (2014) Creating technological innovation via green supply chain management: an empirical analysis. Expert Syst Appl 41(16):6983–6994
62. Wu HH, Chang SY (2015) A case study of using DEMATEL method to identify critical factors in green supply chain management. Appl Math Comput 256:394–403
63. Govindan MK, Devika K, Haq AN (2014) Barriers analysis for green supply chain management implementation in Indian industries using analytic hierarchy process. Int J Prod Econ 147:555–568
64. Kannan D, Khodaverdi R, Olfat L, Jafarian A, Diabat A (2013) Integrated fuzzy multi criteria decision making method and multiobjective programming approach for supplier selection and order allocation in a green supply chain. J Clean Prod 47:355–367. https://doi.org/10.1016/j.jclepro.2013.02.010

65. Govindan K, Rajendran S, Sarkis J, Murugesan P (2015) Multi criteria decision making approaches for green supplier evaluation and selection: a literature review. J Clean Prod 98:66–83. https://doi.org/10.1016/j.jclepro.2013.06.046
66. Wind Y, Saaty TL (1980) Marketing applications of the analytic hierarchy process. Manag Sci 26(7):641–658
67. Bouzon M, Govindan K, Rodriguez CMT (2015) Reducing the extraction of minerals: reverse logistics in the machinery manufacturing industry sector in Brazil using ISM approach. Resour Policy 46(1):27–36
68. Innovation Norway (2018). India EV story emerging opportunities. Retrieved from https://www.innovasjonnorge.no/contentassets/815ebd0568d4490aa91d0b2d5505abe4/india-ev-story.pdf. Accessed 10 Oct 2019.
69. Dunn JB, Gaines L, Kelly JC, James C, Gallagher KG (2015) The significance of Li-ion batteries in electric vehicle life-cycle energy and emissions and recycling's role in its reduction. Energy Environ Sci 8:158–168. https://doi.org/10.1039/C4EE03029J
70. Yang F, Xie Y, Deng Y, Yuan C (2018) Considering battery degradation in life cycle greenhouse gas emission analysis of electric vehicles. Procedia CIRP 69:505–510. https://doi.org/10.1016/j.procir.2017.12.008

Chapter 5
Imports and Exports of Components

5.1 Introduction by Editor

As the EV industry is evolving in India, at present most of the electric and electronics components are imported. The chapter provides some answers regarding the imports of these components. The chapter also examines the exports of component items from India.

5.1.1 Introduction

The objective of this chapter is to examine the importance of imported components in the EV value chain in India. As we have seen in Chap. 2, several critical minerals are imported for batteries. Lithium-ion batteries are also imported. Besides that, other components are also imported.

We divide the EV components into 4 categories:

1. Metal components: Body, chassis, suspension. Gearbox, drive lines, etc. These are all made in India and the supply chain exists for local manufacturing.
2. Electric components: the major components are Battery packs, Motors, chargers, connectors, etc. Most of these components are imported and the supply chain is yet to develop at a large scale in India.
3. Electronics components: Major components are BMS (battery management system), Controller for Motors, CAN controller (controller area networks), and VCU (Vehicle control unit). This category of components is all imported and a supply chain to manufacture in India does not exist at present.
4. Other aggregates: Tyres, glass, lighting. brakes, steering systems, seating, etc. These are all available in India and a large supply chain exists.

S. Ray and D. Mukherji (eds.), *A Primer on Electric Vehicles in India*,
https://doi.org/10.1007/978-981-97-8966-5_5

[1] https://www.wcoomd.org/en/topics/nomenclature/overview/what-is-the-harmonized-system.aspx

5.1.2 Components that Are Imported

We examined the components that are imported by India for the period 2018–19 to 2021–22. The Harmonised System (HS) [1] codes for the list of items for EVs have been taken from Shingal and Aggarwal (2022). From Table 5.1, we can see that electric motors and generators, parts of transformers, etc. are important items that are imported over the years. The value of these items has been increasing (though there are dips in the years 2019–2020 and 2021–22 possibly due to Covid-19.) Other items imported like Lithium-ion and nickel oxide batteries are also imported.

Table 5.1 Components that are imported by India (Value in Rs. lakhs)

HS	Description	2018–19	2019–20	2020–21	2021–22
8708	PARTS AND ACCESSORIES OF THE MOTOR VEHICLES OF HEADINGS 8701 TO 8705	3,292,486 (26)	2,895,485 (25)	2,579,298 (24)	3,601,786 (24)
850,760	LITHIUM-ION	857,408 (7)	881,860.8 (8)	881,145 (8)	1,367,315 (9)
870,899	OTR PRTSANDACCSSRS OF VHCLS OF HDG 8701–8705	1,199,869 (9)	1,035,547 (9)	1,003,881 (9)	1,350,941 (9)
850,440	STATIC CONVERTERS	772,755.5 (6)	766,632.6 (7)	929,959.5 (9)	1,151,122 (8)
8501	ELCTRC MOTRS AND GENRTRS (EXCL GENRTNG SETS)	656,764.1 (5)	634,826.9 (5)	576,370 (5)	811,245.5 (5)
8532	ELCTRCL CAPACITORS FIXD, VARIABLE/ ADJUSTABLE(PRE-SET)	483,163.6 (4)	479,300.4 (4)	489,329.6 (5)	639,328.2 (4)
750,210	NICKEL, NOT ALLOYED	361,350.8 (3)	336,431.4 (3)	313,178.9 (3)	548,389 (4)
Total		10,949,807	10,013,133	9,341,747	12,961,081

Source Ministry of Commerce and Industries, Government of India
Note The figure in parentheses is the shares in the imports of EV-related items in that year

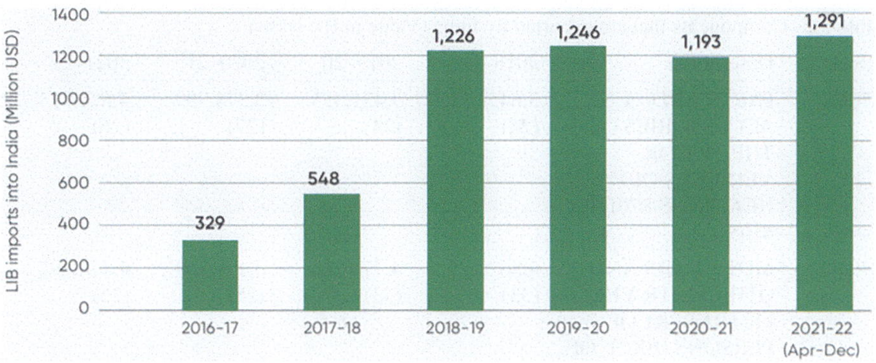

Fig. 5.1 Lithium-Ion batteries imported into India [2]

5.1.3 Sourcing of Components

India imported lithium-ion batteries worth $ 1.23 billion in 2018–19 (Gupta, 2020). Of this, $ 773 million was from China, $ 267 was from Hong Kong and $114 million was from Viet Nam. Cell manufacturing involves a very complex supply chain from mineral extraction to battery pack making. Due to the geographic location of most of the lithium, and cobalt mines in Latin America, Africa, Australia, and nickel in Indonesia. After extraction, the materials travel to China for refining. After refining the materials have to travel back to Europe or USA for cell manufacturing and thereafter pack and finally end up in the EV. Battery manufacturing in the world is dominated by CATL (China), LG (South Korea) and Panasonic (Japan).

From Fig. 5.1, we also note that the imports of lithium-ion batteries are increasing rapidly.

5.1.4 Exports

To set off imports, certain exports do happen from India. These are largely in two categories: Motor vehicles and motor vehicle components, parts, accessories etc.

From Table 5.2 we note that parts and accessories of the motor vehicles are the top most exported items [3]. Also, motor cars and other motor vehicles for the transport of persons are also an important export item; although their importance has been declining. The value of these items has been increasing (though there are dips in the years 2019–2020 and 2021–22 possibly due to Covid-19). In 2021–22, we see a rebound in the volume. We also note that some of the items that are exported like gearboxes, electric motors, generators, static converters, etc. are also imported (e.g. items with HS 8708, 870,899, 850,440 etc.). This could be an indication of integration in value chains. One has to examine the items at the 8-digit level to get a better understanding.

Table 5.2 Components that are exported by India (Value in Rs. lakhs)

HS	Description	2018–19	2019–20	2020–21	2021–22
8708	PARTS AND ACCESSORIES OF THE MOTOR VEHICLES OF HEADINGS 8701 TO 8705	3,645,931 (24)	3,421,355 (24)	3,374,989 (27)	4,670,373 (26)
8703	MOTR CARS AND OTHR MOTR VHCLS FR TRNSPRT OF PERSONS (EXCL OF 8702) INCL RCNG CARS ETC	4,862,124 (33)	4,774,536 (33)	3,179,209 (25)	4,420,492 (25)
870,899	OTR PRTSANDACCSSRS OF VHCLS OF HDG 8701–8705	1,971,003 (13)	1,749,618 (12)	1,740,993 (14)	2,272,452 (13)
8704	MOTOR VEHICLES FOR THE TRANSPORT OF GOODS	931,189.8 (6)	611,392.5 (4)	517,941.7 (4)	1,096,142 (6)
850,440	STATIC CONVERTERS	565,388.6 (4)	881,586.8 (6)	794,545.3 (6)	1,009,937 (6)
8714	PRTS AND ACCSSRS OF VHCLS OF HDG 8711–8713	440,365.4 (3)	462,888.9 (3)	522,205.1 (4)	733,447.3 (4)
871,410	OF MOTORCYCLES (INCLUDING MOPEDS):	251,606.9 (2)	277,820.3 (2)	347,326.8 (3)	508,915.7 (3)
Total		13,478,433	13,073,073	11,365,612	15,862,143

Source Ministry of Commerce and Industries, Government of India
Note The figure in parentheses is the shares in the exports of EV-related items in that year

[2] IESA

[3] These items account for almost 90% of the items that are exported among the EV items using the codes provided by Shingal and Aggarwal (2022).

Conclusion

At present India's electric vehicle supply chain is heavily dependent upon the import of raw materials and components. Most of the raw material imports are from China as it controls the majority of battery materials like Lithium, Cobalt, Nickel, Manganese, etc. Many parts are also imported as we saw in this chapter. How India can move up the electric vehicle value chain by manufacturing these parts remains to be seen.

Machine Generated

Parlaktas et al. (2019) present a novel approach for the design and manufacturing of an electric vehicle chassis, keeping in mind manufacturing simplicity, lightness, robustness, and low cost. Total inventory and material costs can be kept very low with a lightweight and low-cost chassis built from simple planar parts that can also perform very well. Shenoy (2016) says India is taking the right path toward electric mobility solutions in terms of policy, preparedness, and capabilities and increased promotion of renewable energy, will help solve the energy crisis. Jerome and Udayakumar (2021) investigate the possible savings in gasoline consumption and reduction in CO_2 emission in India from the battery-electric two-wheeler segment during the fiscal year 2021–2030. Gupta et al. (2022) study the risk factors associated with the Public Private Partnership project of EVs across India. The risk factors are identified, with the help of a literature review and inputs received from industry experts, and are in four main categories - financial, market, political/legal, and operational risks. Liu et al. (2019) use a combination of a series of computer-aided simulation analysis techniques, to develop a body structure for the US market on the basis of the required domestic body structure. Mall and Ambanandam (2022) use the fuzzy-AHP MCDM technique for the criteria weight calculation for ranking alternatives in the study. They conclude that the option of battery swapping is the best out of all the available alternatives.

References

IESA News (n.d.) India set to acquire lithium for EV batteries in 6 months. Link: https://indiaesa.info/news-menu/1903-india-set-to-acquire-lithium-for-ev-batteries-in-6-months

Khanij Bidesh India Limited (KABIL) (n.d.) Ministry of Mines. Link: https://mines.gov.in/webportal/content/kabil

Shingal, A. and Agrawal, P. (2022) Trade and tariffs along the EV value-chain: literature review and stylized facts. Chapter 4. In Moerenhout, T., Goldar, A., Ray, S., Shingal, A., Goel, S., Agarwal, P., Jain, S., Thakur, V., Tarun., Gaurav, K., Aneja, DA., Bansal, A. "Understanding Investment, Trade, and Battery Waste Management Linkages for a Globally Competitive EV Manufacturing Sector." Report. New Delhi. 2022. https://icrier.org/publications/understanding-investment-trade-and-battery-waste-management-linkages-for-a-globally-competitive-ev-manufacturing-sector/

World Customs Organisation (n.d.) What is the Harmonized System (HS)? Link: https://www.wcoomd.org/en/topics/nomenclature/overview/what-is-the-harmonized-system.aspx

Footnotes/Endnotes

[1] https://www.wcoomd.org/en/topics/nomenclature/overview/what-is-the-harmonized-system.aspx

[2] IESA

[3] These items account for almost 90% of the items that are exported among the EV items using the codes provided by Shingal and Aggarwal (2022).

5.2 Machine Generated Summaries

Disclaimer: The summaries in this chapter were generated from Springer Nature publications using extractive AI auto-summarization: An extraction-based summarizer aims to identify the most important sentences of a text using an algorithm and uses those original sentences to create the auto-summary (unlike generative AI). As the constituted sentences are machine selected, they may not fully reflect the body of the work, so we strongly advise that the original content is read and cited. The auto generated summaries were curated by the editor to meet Springer Nature publication standards. To cite this content, please refer to the original papers.

Machine generated keywords: weight, twowheeler, india, design, crash, availability, fuzzy, passenger, safety, mass, co, body, manufacturing, co emission, charge.

5.2.1 The Design and Manufacturing Process of an Electric Sport Car (EVT S1) Chassis [1]

This is a machine-generated summary of:

Parlaktaş, Volkan; Tanık, Engin; Babaarslan, Nahit; Çalık, Göktuğ Burak: The Design and Manufacturing Process of an Electric Sport Car (EVT S1) Chassis [1].

Published in: Iranian Journal of Science and Technology, Transactions of Mechanical Engineering (2019).

Link to original: https://doi.org/10.1007/s40997-019-00328-6.

If you want to cite the papers, please refer to the original.

For technical reasons we could not place the page where the original quote is coming from.

Abstract-Summary

"This study presents novel approaches for design and manufacturing of an electric vehicle chassis and its fixtures with very low cost."

"The chassis is tub type and its original structure is formed regarding manufacturing simplicity, lightness, robustness, and low cost."

"The design philosophy of the chassis is discussed in detail."

"Original fixture designs for low-cost manufacturing are introduced in this study."

"Manufacturing cost of the chassis, fixtures, and the vehicle is presented."

"It is evidenced that a lightweight and low-cost chassis built from simple planar parts may perform very well."

"Electric vehicles possessing such simple and robust chassis may be popular due to the manufacturing advantages against today's extremely complex designs."

Introduction

"Even lithium-based batteries possess a relatively low specific energy; therefore, heavy vehicles require a large quantity of batteries (Rao and others [2]; Cui and Zhang [3]; Baek and others [4])."

"Lithium-based batteries are still pricey; therefore, EVs cannot compete with internal combustion engine vehicles (ICEVs) when cost is taken into account (Chan [5])."

"For an EV, designing a low-weight chassis and components is crucial as shown by earlier studies (Leitman and Brant [6]; Fenton and Hodkinson [7]; Chan and Chau [8])."

"The main objective is to build a light and robust EV chassis with low cost."

"Design and manufacturing process of low-cost novel fixtures of the chassis are presented."

"Manufacturing procedures for low-cost production are explained, and cost analysis of the chassis, fixtures, and the car is listed."

"After all of the manufacturing and assembly stages, the EV is at the state of running and it is called as EVT S1 (Electric Vehicle, Turkey)."

Design Philosophy

"The chassis type is the main factor that increases the cost."

"Cost of the tools used for manufacturing a unibody chassis is very high."

"Investment cost of tub chassis manufacturing is relatively small."

"Aluminum is referred as the optimum vehicle chassis material when cost and lightness are both taken into account (Larminie and Lowry [9])."

"We also preferred aluminum as the main material of the chassis."

"The novelty in our design philosophy is to build a chassis that is formed by planar and simple aluminum profiles that are readily available in the stocks of producers."

"All of these modules are designed considering numerous parameters such as crash safety, torsional rigidity, ergonomics, location of motor, battery, and other chassis components."

"It is essential to design the chassis by considering feasibility of the fixtures to keep cost low."

Rigidity and Impact Analyses

"That this is the torsional rigidity value of the chassis solely."

"The major concern during the design is that the front subframe should not be very resistant; it should be a crumple zone that absorbs the kinetic energy while deforming, adversely the deformation of the safety cell (tub module in tub-type chassis) should be as small as possible."

"The selected steel is very ductile, but also strong enough (yield strength = 700 MPa); hence, it prevents the profiles of the front subframe to tear out the front wall and penetrate into the safety cell in the case of extreme impact forces."

"We repeated this simulation with a speed of 100 km/h. The deformation of the safety cell was still small, whereas the maximum acceleration of this crash was 75 g. The same simulations are also adopted for backwards crash and similar results are obtained."

Original Fixture Design

"Preciseness of the chassis can be satisfied by manufacturing it with the aid of fixtures."

"As the chassis modules consist of planar parts and they are parallel to x–y plane, it is possible to implement a planar-shaped main fixture."

"There are grooves on this planar fixture with different dimensions, where the aluminum profiles of the chassis fit in."

"If a profile is manufactured in lower tolerance limit, there is a clearance fit between the profile and its groove in the fixture."

"The profiles at the bottom of subframe fit in the grooves of the main fixture."

"This fixture is manufactured from three different steel plates with a two-axis laser CNC."

"The door chassis that is integrated to the main chassis should also be manufactured precisely."

"Preciseness can only be satisfied by manufacturing the door chassis with a fixture as well."

Manufacturing Process

"When the design of main parts of the chassis is completed, all of the parts are assembled together in Catia."

"Consistencies of the moving parts are checked in this way before manufacturing process."

"If there exists an error, money and time waste are inevitable during manufacturing, so all of the design steps should be checked carefully."

"According to our operation sequence, first, parts of the tub module are fixed on the fixture and parts that are contacting with large surfaces are assembled with a special adhesive."

"During locating the chassis profiles on the main fixture platform, shims are used between grooves of the fixture and profiles to obtain precise tolerances."

"The complete front and rear subframes are connected to the tub module with aluminum bolts and nuts in order to obtain a detachable assembly."

"Assembling all of the modules on the same fixture platform minimizes the cumulative errors in tolerances."

Cost Analyses

"Low-cost production ability is one of the most important novelties of this chassis."

"For mass production these costs would also decrease, because it is well known that when the amount of order increases, the cost decreases."

"Table, it can be concluded that, for a single chassis, the cost would be less than 975$ during mass production, excluding the labor (varies too much depending on the country)."

"Our manufacturing procedures do not require too much workmanship for an aluminum tub-type chassis."

Test Results

"The chassis should be checked attentively against a catastrophic failure."

"It was observed that there was no indication of failure in suspension components and chassis."

"Torsional rigidity was tested by climbing to high pavements: While only one of the wheels of the car was on a high pavement, we observed that the amount of twist of the chassis was very small; no squeak could be heard and the doors can be opened and closed just like the car is on a flat surface."

Conclusions

"General design philosophy of an original EV chassis is presented."

"Design and manufacturing methodology of the original fixtures with very low cost are shown."

"It is concluded that total inventory and material cost can be kept very low with the design philosophy introduced in this study."

"With this study, it is shown that a lightweight and low-cost chassis built from simple planar parts can also perform very well."

5.2.2 Electric Mobility in India—A Bosch Perspective [10]

This is a machine-generated summary of:

Shenoy, R K: Electric Mobility in India—A Bosch Perspective [10].

Published in: Auto Tech Review (2016).

Link to original: https://doi.org/10.1365/s40112-016-1089-8.

If you want to cite the papers, please refer to the original.

For technical reasons we could not place the page where the original quote is coming from.

Introduction

"India is one of the fastest growing automobile markets in the world today, thanks to increasing income levels of the middle class."

"In the government's scheme of things, some broad categories considered include electric vehicles (EV), hybrid electric vehicles, plug-In hybrid vehicles, and fuel cell vehicles."

"Some specific advantages that electric drive technologies have over traditional international combustion (IC) engines, apart from decreased emissions, include the following::: Electric motors are inherently more efficient than combustion engines, effectively utilising more than 90% of the energy provided, compared to the 37% in IC engines [11]."

"Continuous acceleration and braking is very bad for IC engines."

"Electric motors, on the other hand, are very efficient in such conditions because of high torque generation and re-generation, while braking."

Market Assessment

"The electric vehicle and HEV market in India is nascent but growing."

"With the growing interest in cleaner lifestyle and growing income of the society, electric vehicle manufactures can see substantial growth if the market, is properly monetised."

"According to the latest market reports, the total number of electric two-wheelers on Indian roads stands at about 400,000 units, and electric passenger cars account for about 1,500 units [12]."

Challenges

"In terms of technology availability and affordability, we are far behind global markets."

"There are quite a few primary challenges for electrification in India, including the following::: Lack of definite guidelines & regulations, and difficulties in physical implementation;:: EV charging infrastructure and quality/ availability of grid power;:: Low domestic availability of advanced electronics and controls;:: Lack of cost-effective battery technology;:: Large investments are needed, but stakeholders perceive the area as one with high risk; and:: Pricing of the vehicles and consumer awareness."

Government Initiative

"The Indian government has introduced various schemes and incentives for promotion of electric vehicles."

"Unveiled in January 2013, the National Electric Mobility Mission Plan 2020 (NEMMP) lays out a detailed and comprehensive roadmap for electrification of the road transport sector in India."

"It aims to grow sales of electric vehicles to six-seven million units a year by 2020 and that includes direct subsidies for the purchase of both four- and two-wheeled electric vehicles."

"To promote eco-friendly vehicles, the government launched the FAME [Faster Adoption and Manufacturing of (Hybrid &) Electric vehicles] India scheme offering incentives on electric and hybrid vehicles of up to ₹ 29,000 for two-wheelers, and ₹ 1.38 lakh for cars."

Round-Up

"With the increased promotion of renewable energy, we will not just solve the energy crisis, but also increase the percentage of green energy."

"India is taking the right path towards electric mobility solutions in terms of policy, preparedness and capabilities."

"With this, we will be able to give our children the same fresh air to breathe in metros like Delhi and Bangalore, which we had respired in smaller towns earlier."

5.2.3 Assessment of Gasoline Consumption and Greenhouse Gas Emission Reduction on Using Battery Electric Two-Wheeler in India [13]

This is a machine-generated summary of:

Jerome, Sajan; Udayakumar, M.: Assessment of gasoline consumption and greenhouse gas emission reduction on using battery electric two-wheeler in India [13].

Published in: Journal of the Brazilian Society of Mechanical Sciences and Engineering (2021).

Link to original: https://doi.org/10.1007/s40430-020-02756-x.

If you want to cite the papers, please refer to the original.

For technical reasons we could not place the page where the original quote is coming from.

Abstract-Summary

"This study investigates the possible savings in the gasoline consumption and reduction in CO_2 emission in India from the battery electric two-wheeler segment during the fiscal year 2021–2030."

"Any policy actions intended at the reduction in CO_2 emission and gasoline consumption from the transportation sector must focus on the two-wheeler segment."

"Older two-wheeler has higher gasoline consumption due to lower fuel economy which leads to larger CO_2 emission."

"The study reveals that with 100% replacement of newly registered ICE two-wheelers with BE two-wheelers, a total savings of 322.50 billion litres of gasoline and a reduction of 571.49 million tonnes of CO_2 emissions can be achieved during the period 2021–2030."

"With strict implementation of a scrapping policy whereby all ICE two-wheelers completing 15 years of service are taken off the road, the gasoline savings and reduction in CO_2 emissions can be enhanced to 403.23 billion litres and 811.27 million tonnes, respectively, during the same period."

Introduction

"Battery electric vehicles store electrical energy on an on-board battery and propel the vehicle with the help of an electric motor and hence they produce no emissions at the point of use."

"India is not an exception with more than 78% of the electricity produced in the year 2019 came from fossil fuel-based power plants [14, 15]."

"The effect of the use of BE two-wheeler in CO_2 mitigation will depend on the extent of the contribution from non-fossil fuel-based energy sources for power generation."

"The promotion of electric vehicles with the switch to non-fossil fuel-based power generation can reduce India's import dependence on crude oil and can lead to increased consumption of renewables like solar and wind [16]."

"The analysis reveals that the substitution of conventional two-wheelers with BE two-wheelers can lead to a huge reduction in the amount of GHG emissions in the future with greater penetration of non-fossil fuel-based renewable sources into the generation mix."

Methodology

"We develop a simple model to estimate the CO_2 emission and the gasoline consumption based on emission coefficients, average fuel consumption and the annual mileage."

"The reliance on fossil fuels in the transportation segment can be reduced by the use of EVs, if the electricity needed for their propulsion is sourced from non-fossil fuel-based energy sources such as nuclear, hydro and other renewable sources [17]."

"The future electrical power structure in India can be predicted by considering the drifts in the historical variations in the installed capacity of different sources of energy."

"In India, the largest average utilization hour is obtained for nuclear power (5908.82 h/year), followed by coal and lignite (4984.35 h/year)."

"In the year 2030, the contribution from non-fossil fuel-based energy, which includes hydro, nuclear, wind, solar, bio-power and small hydro, approaches close to 40%, which is in unison with the Intended Nationally Determined Contribution (INDC) [18]."

Results and Discussion

"With the existing electricity generation mix of India in 2019, the CO_2 emitted per kWh of electricity produced at power plants is computed as 726 g. With 100% replacement of all the newly registered two-wheelers with BEL1 two-wheelers, the CO_2 emission in 2030 will be limited to 201.89 million tonnes when compared with 322.83 million tonnes without any replacement, leading to a reduction of 120.94 million tonnes."

"This section describes the annual savings in gasoline consumption and reduction in CO_2 emission that can be achieved over a decadal period on scrapping all the ICE two-wheelers that are completing 15 years on road."

"In the case of the BEL2 category which consumes more electrical energy in the order of 0.06 kWh/km, the annual CO_2 emission for the same years will be more in the range of 381.58, 352.67, 322.72 and 303.27 kg, resulting in a percentage decrease of 19.33, 25.44, 31.77 and 35.88% when compared with the conventional ICE two-wheeler."

Conclusion

"This study analysed the savings in gasoline consumption and reduction in CO_2 emission related to the transportation sector in India on replacing the conventional ICE two-wheelers with BE two-wheelers."

"With strict implementation of a scrapping policy whereby all ICE two-wheelers completing 15 years of service are taken off the road, the gasoline savings and reduction in CO_2 emissions can be enhanced to 403.23 billion litres and 811.27 million tonnes, respectively, during the same period."

"The enhanced use of non-fossil fuel-based sources for electricity generation will lead to a further reduction in CO_2 emission, as the energy required for the propulsion of BE two-wheeler is sourced from the grid."

"The enhanced contribution from the non-fossil fuel-based energy source leads to lower emission rates per unit electricity generated, and hence a noticeable reduction in CO_2 emission from the battery-operated electric two-wheeler which sources the power from the grid."

5.2.4 Risk Evaluation of Electric Vehicle Charging Infrastructure Using Fuzzy AHP—A Case Study in India [19]

This is a machine-generated summary of:

Gupta, Shubham; Khanna, Raghav; Kohli, Pranay; Agnihotri, Sarthak; Soni, Umang; Asjad, M.: Risk evaluation of electric vehicle charging infrastructure using Fuzzy AHP—a case study in India [19].

Published in: Operations Management Research (2022).

Link to original: https://doi.org/10.1007/s12063-022-00290-8.

If you want to cite the papers, please refer to the original.

For technical reasons we could not place the page where the original quote is coming from.

Abstract-Summary

"Wide adoption of Electric Vehicles (EV) may require an effective and efficient network of charging infrastructure across the nation."

"Work an attempt has been made to study and analyze the risk factors associated with Public Private Partnership Project of EV across the India."

"The risk factors are identified, with the help of literature review and inputs received from the industry experts, and are divided into four main categories which consist of financial, market, political/legal and operational risks."

"The proposed research work will be helpful for the practicing engineers, academician, researcher and the concern Industries in order to promote the EV charging infrastructure in the most efficient and effective manner."

Introduction

"EV adoption in India has been faster in the mass mobility sector, with the adoption of E-Rickshaws and electric buses, than in the personal mobility segment, but the growth of EVs in India is still lower than projected."

"In order to address these issues, the Government can focuses on the PPP model to increase its usage and adoptability."

"A numerous risk factors are involved in adoption and implementation of PPP model especially for charging infrastructure, although there are extensive research reported in the literature but most of them are fail to address the risk involved in establishing the EVCI through PPP model (Awasthi and others [20]; Bing and others [21]; Sheppard and others [22]; Thomas and others [23])."

"Among the different tools and techniques Analytical Hierarchy Process (AHP) is one of the oldest techniques that are widely applicable and adopted in the numerous research work (Wang and Elhag [24])."

Literature Review

"The Multi Criteria Decision Making (MCDM) techniques are widely applicable to study and implement the contradicting, multi-dimensional, uncertain and incommensurable effects of various decisions explicitly (Awasthi and Omrani [25])."

"This is because AHP technique involves subjectivity in pairwise comparisons and the decision-makers may not be able to assign the desired crisp values to comparison judgments (Chan and Kumar [26]; Yrjölä [27])."

"Studies focused on providing a comprehensive study on risk assessment on charging infrastructure in Indian context are limited in literature, especially in case of PPP model."

"Studies pertaining to the application of hybrid MCDM techniques on ranking the risk factors involved in gauging the charging infrastructure are scares."

"A very few studies are reported on the application of sensitivity analysis on hybrid MCDM technique in case of risk assessment of charging infrastructure."

Methodology

"In order to make an electric vehicles become more sustainable, it is required to identify the risk factors involved in the charging infrastructure of these vehicles and then adopt a methodology to identify the most critical one."

"Step 2: Assigning Weights to Judgement Matrices The judgement matrix TS tells us about the pairwise comparison of the project at different stages, and TRS denotes the judgement matrix of risk factors under a particular stage St. The method used to compute the weighting of fuzzy matrices is the Geometric Mean Technique by using fuzzy arithmetic operation Step 3: Defuzzifying the assigned Weights Defuzzifying is done by using the α-cut method, the value of α depends on the basis of experts' familiarity with the project."

"Risk index μ is considered to get crisp weightings by incorporating the expert's attitude towards risk factors."

Results and Discussion

"Political Risks - These are the risks faced by the PPP projects from public sectors including the current government, opposition parties and countries' legislature."

"After calculation, the results show Government Policies (R31) as the most dominant risk factor among all in this project."

"Democratic nations like India have a highly uncertain political situation which makes Foreign direct investment (R33) 3rd most important risk factor."

"This makes Corruption (R32) 4th most important risk factor for this project."

"This make competitor risk (R21) 5th most important risk factor."

"Variance based sensitivity analysis can be carried out by assuming the independence of the expert judgment pertaining to risk evaluation of charging infrastructure."

"A feasible range of deviations in the risk evaluation of charging infrastructure attributes is specified in terms of a bounded set of discrete percent change of 10% from the original attribute value (i.e. Expert Judgment) generally used for the analysis."

Conclusion

"This paper aims to point out the risks involved that may influence the PPP project and look at it with a mathematical approach that may help the private as well as the government sector."

"A framework was created to identify and evaluate the risks involved in a PPP EVCI project with an integrated fuzzy AHP approach."

"Government Policies', R31, emerges as the most dominant risk factor as the Govt."

"Risk analysis is a part of main infrastructure construction and hence, plays a significant role."

"The concept of Electric vehicle charging infrastructure under a PPP project is pretty unique and new in India, the risk factors identified can't be perfect."

5.2.5 Innovative Electric Vehicle Body Design Based on Insurance Institute for Highway Safety Side Impact Conditions [28]

This is a machine-generated summary of:

Liu, Xinchun; Liang, Maoyan; Luo, Qiang: Innovative Electric Vehicle Body Design Based on Insurance Institute for Highway Safety Side Impact Conditions [28].

Published in: Automotive Innovation (2019).

Link to original: https://doi.org/10.1007/s42154-019-00070-4.

If you want to cite the papers, please refer to the original.

For technical reasons we could not place the page where the original quote is coming from.

Abstract-Summary

"A version of an electric vehicle was developed and designed for the US market on the basis of the required domestic body structure."

"Using a combination of butterfly diagram analysis, bending moment management, section forces and other computer-aided simulation and analysis techniques, this paper presents a body structure design that can achieve a "GOOD" evaluation under the US Insurance Institute for Highway Safety (IIHS) side impact body structure conditions by optimizing the force transfer path, the B-pillar deformation mode and the threshold support structure."

"The threshold support structure supports realization of the "GOOD" rating for IIHS side impact and helps the body to meet the crash requirements of the Federal Motor Vehicle Safety Standard FMVSS214 and the US New Car Assessment Program (NCAP) requirements for side impact at 32 km/h and 75° angular pole impact."

Introduction

"Among the development steps, the design of the deformation structure serves as a foundation to improve the vehicle body's safety performance in terms of car crash safety."

"With the continuing popularity of passenger cars in the global market, the environmental problems caused by traditionally fueled vehicle emissions are becoming increasingly serious [29], so many countries are also actively developing new energy-type vehicles."

"This means that the body structures of electric vehicles must be able to withstand greater forces and absorb more kinetic energy to improve their safety."

"An electric vehicle developed for the US market has a high positioning and is required to achieve the highest evaluation of the US market regulations and standards."

"It is necessary to design an innovative body structure that can meet the relevant requirements of electric vehicles while meeting the highest evaluation under the US standards."

Objectives and Requirements

"The basic domestic vehicle is a compact sport utility vehicle (SUV) with five seats, a curb weight of 1650 kg and conventional fuel engine as a power system."

"The new model is a purely electric vehicle based on the basic domestic model with a 500 km range."

"The requirement for this new model is to develop an electric vehicle that can satisfy the Federal Motor Vehicle Safety Standards (FMVSS) regulations [30] for the US market and can also achieve a five-star evaluation from the US-NCAP [31] and the "PICK+" evaluation from the IIHS Top Safety ratings [32]."

Design Optimization Method and Process

"The impact's kinetic energy can only be absorbed through small amounts of deformation of the B-pillar, the side door anti-collision beam, the floor beam, the roof cross-rail and related structures."

"On the body side, the deformation can be restrained by optimizing the force transfer path and the structural deformation mode and improving the strengths of the key parts and joints of the body to reduce the intrusion volume and speed for the side doors and the B-pillars."

"The overturning of the threshold leads to zero deformation in the area below the buttock level of the B-pillar, which thus wastes precious space that can absorb energy without harming the passengers during a side impact."

"Through optimization of force transfer path, B-pillar deformation mode and threshold support structure, the body intrusion under the IIHS side impact conditions has improved significantly."

Conclusions

"Using a combination of a series of computer-aided simulation analysis techniques, a body structure has been designed that can achieve a "GOOD" evaluation in terms of the IIHS side impact body structure conditions in the USA."

"All the preset objectives are achieved via the design optimization, and the body structure is optimized to change from a "POOR" evaluation to a "GOOD" evaluation under the IIHS side impact conditions."

"This "GOOD" evaluation of the company's electric vehicle body structures under IIHS side impact conditions has achieved a breakthrough from scratch."

"A new purely electric body structure is designed that will lay a foundation for more models to achieve a "GOOD" evaluation for their IIHS side impact body structures in the future."

5.2.6 A Fuzzy Analytic Hierarchy Process and VIKOR Framework for Evaluation and Selection of Electric Vehicle Charging Technology for India [33]

This is a machine-generated summary of:

Mall, Shaurya; Anbanandam, Ramesh: A Fuzzy Analytic Hierarchy Process and VIKOR Framework for Evaluation and Selection of Electric Vehicle Charging Technology for India [33].

Published in: Transportation in Developing Economies (2022).

Link to original: https://doi.org/10.1007/s40890-022-00150-x.

If you want to cite the papers, please refer to the original.

For technical reasons we could not place the page where the original quote is coming from.

Abstract-Summary

"The climatic and environmental issues, rapid urbanisation, and technological developments in renewable energy in recent years have led to the rise of electric vehicles (EVs) for sustainable transportation."

"The serviceability of a charging station is greatly affected by its location and the technology available."

"To fully incorporate EVs in the transportation sector, their charging technologies and stationing at appropriate charging sites must be analysed thoroughly."

"The selection of electric vehicle charging technology is a complex procedure."

"It can be decomposed into a decision-making problem with various criteria and alternatives to be considered simultaneously."

"This can be done using a multi-criteria decision-making (MCDM) technique."

"This study provides a fuzzy analytic hierarchy process (Fuzzy-AHP) framework to evaluate different criteria affecting our alternate technologies."

"Further, another MCDM Technique, VIKOR, is used to find solutions closest to the ideal condition and rank the alternate technologies."

Introduction

"It is widely accepted that Electrical vehicles (EVs) have helped lower the tailpipe emission."

"The Government promotes the adoption of EVs by running projects such as faster adoption and manufacturing of hybrid and electric vehicles (FAME) Phase I scheme, 2015, and Phase II under the National Electric Mobility Mission Plan (NEMMP), 2013, initiative [34]."

"In EV 30@30 campaign, which aims to reach 30% market shares for electric vehicles in all modes except for two-wheelers by the year 2030, the projected share of EVs in 2030 in India is 29% [35]."

"The availability of charging infrastructure is one of the crucial barriers other than the driving range of electric vehicles, which mainly affects the adoption [36]."

"Charging stations' availability is crucial for facilitating EV mass adoption."

"To incorporate Electric Vehicles on a full-scale in the transportation sector in India, their charging technologies and stationing at appropriate charging sites need to be analysed thoroughly."

Literature Review

"Wu and others [37] has used PROMETHEE and ANP to measure the correlation of the indicators in electric vehicle charging stations site selection."

"From the available literature, we have identified the following alternatives technologies employed for charging EVs."

"There are two types of charging under conductive charging, i.e., AC and DC, which can be further classified under three levels."

"The 'EV charging Level, I' vehicle uses a board charger that works with the existing AC supply network."

"In' EV charging level II', electric vehicle supply equipment is used to charge the vehicle using an AC power supply."

"Inductive charging is expensive and slower than the available conductive charging technologies."

"Battery Swapping is a tariff-based charging service provided by the manufacturer company."

"of India has already prescribed the standards for safety and performance for AC charging stations for EV and HEV application in Indian conditions."

Methodology

"We will be using the fuzzy-AHP MCDM technique for the criteria weight calculation instead of AHP in this study."

"This problem can be tackled by incorporating fuzzy theory into AHP to solve multi-criteria decision-making problems [38]."

"The boundaries l, m and n indicate the smallest possible value, the most promising value and the largest possible value of a fuzzy set, respectively."

"[39] extent analysis method is used for criteria prioritisation in Fuzzy AHP."

"The triangular fuzzy number can be denoted by (l, m, u)."

"When l = m = u, is a non-fuzzy number by convention."

"We can get 'k' extent analysis values for each object, with the following signs:where all the M_{gi}^{j} (j = 1, 2, ……, k) are fuzzy triangular numbers."

"Since M1 and M2 are convex fuzzy numbers, we have that."

Case Study

"We have identified three potential alternatives to plan the ideal charging technology, keeping Delhi as a reference."

"All these alternatives are identified from literature keeping the Indian urban environment in mind."

"Each expert was asked to compare the criteria based on the Saaty scale of 1–9, where 1 represents the equal importance between the compared criteria and 9 represents absolute relative importance [40]."

"The weights by the experts were given according to the following table: The data consistency is checked according to the procedure suggested by Saaty [40]."

Results and Discussion

"This study uses [39] extent analysis for the weight calculation of the alternatives."

"Based on these best and worst cases, the distance of each alternative from the ideal positive Pj and from the negative ideal Nj is calculated."

"The ranking of the alternates is done based on Qi value."

"The change in the qi score of all the alternatives due to variation in the value of V, is noted."

"From the sensitivity analysis, it is illustrated that the value of Qi is lowest for alternative 3 for all the values of v. Hence the optimal option is alternative 3, i.e., the Battery swapping method."

Conclusion

"As the development of charging infrastructure and deployment of smart grid technologies are at their initial phase, this study aims to identify the best charging technology selection from the available alternatives based on the seven criteria specified in the literature."

"This study can be helpful for the policymakers to plan for the possible charging technology options in the urban environment."

"This study has prepared a fuzzy-AHP and VIKOR framework to evaluate the criteria and rank the alternatives."

"From the findings of this study, the option of battery swapping is the best out of all the available alternates."

"This study can further be extended by including different criteria or sub-criteria in the evaluation framework or any other MCDM method such as ANP."

"There may be more alternatives available in the future for charging with technological advancements, which are not considered in this study."

Bibliography

1. Parlaktaş V, Tanık E, Babaarslan N, Çalık GB (2019) The design and manufacturing process of an electric sport car (EVT S1) chassis. Iranian J Sci Technol Trans Mechan Eng. https://doi.org/10.1007/s40997-019-00328-6
2. Rao GS, Rao GK, Raju SSN (2014) An innovative approach to battery management and propulsion system of electric/hybrid electric vehicle. Int J Electric Hybrid Veh 6:1–13
3. Cui JY, Zhang XW (2015) Two-layer distributed equalization management system for electric vehicle power battery. Int J Automot Technol 16:1007–1016
4. Baek KW, Hong ES, Cha SW (2015) Capacity fade modeling of a lithium-ion battery for electric vehicles. Int J Automot Technol 16:309–315
5. Chan C (2002) The state of the art of electric and hybrid vehicles. Proc IEEE 90:247–275
6. Leitman S, Brant B (2008) Build your own electric vehicle. McGraw-Hill Professional, New York
7. Fenton J, Hodkinson R (2001) Lightweight electric/hybrid vehicle design. Butterworth-Heinemann, Oxford
8. Chan CC, Chau KT (2001) Modern electric vehicle technology. Oxford University Press, Oxford
9. Larminie J, Lowry J (2003) Electric vehicle technology explained. Wiley, New York
10. Shenoy RK (2016) Electric mobility in India—a Bosch perspective. Auto Tech Rev. https://doi.org/10.1365/s40112-016-1089-8
11. Sperling & Gordon (2009)
12. Source: News Report, Cars21.com
13. Jerome S, Udayakumar M (2021) Assessment of gasoline consumption and greenhouse gas emission reduction on using battery electric two-wheeler in India. J Brazilian Soc Mechan Sci Eng. https://doi.org/10.1007/s40430-020-02756-x
14. CEA-Central Electricity Authority (2019) Ministry of Power, Government of India. CEA Annual Report 2018–19. http://cea.nic.in/reports/annual/annualreports/annual_report-2019.pdf. Accessed 01 Mar 2020
15. CEA-Central Electricity Authority (2019) Ministry of power, Government of India. Growth of electricity sector in India from 1947–2019. http://www.cea.nic.in/reports/others/planning/pdm/growth_2019.pdf. Accessed 01 Mar 2020
16. Gupta D, Ghersi F, Vishwanathan SS, Garg A (2019) Achieving sustainable development in India along low carbon pathways: macroeconomic assessment. World Dev 123:104623
17. Tran M, Banister D, Bishop JD, McCulloch MD (2013) Simulating early adoption of alternative fuel vehicles for sustainability. Technol Forecast Soc Chang 80(5):865–875
18. MNRE-Ministry of New and Renewable Energy (2017) Government of India, annual report 2016–17. https://mnre.gov.in/knowledge-center/reports. Accessed 01 Mar 2020
19. Gupta S, Khanna R, Kohli P, Agnihotri S, Soni U, Asjad M (2022) Risk evaluation of electric vehicle charging infrastructure using Fuzzy AHP—a case study in India. Operations Management Research. https://doi.org/10.1007/s12063-022-00290-8
20. Awasthi A, Venkitusamy K, Padmanaban S, Selvamuthukumaran R, Blaabjerg F, Singh AK (2017) Optimal planning of electric vehicle charging station at the distribution system using hybrid optimization algorithm. Energy 133:70–78

21. Bing L, Akintoye A, Edwards PJ, Hardcastle C (2005) The allocation of risk in PPP/PFI construction projects in the UK. Int J Project Manage 23(1):25–35
22. Sheppard CJ, Gopal AR, Harris A, Jacobson A (2016) Cost-effective electric vehicle charging infrastructure siting for Delhi. Environ Res Lett 11(6):064010
23. Thomas AV, Kalidindi SN, Ananthanarayanan KABT (2003) Risk perception analysis of BOT road project participants in India. Constr Manag Econ 21(4):393–407
24. Wang YM, Elhag TM (2006) Fuzzy TOPSIS method based on alpha level sets with an application to bridge risk assessment. Expert Syst Appl 31(2):309–319
25. Awasthi A, Omrani H (2019) A goal-oriented approach based on fuzzy axiomatic design for sustainable mobility project selection. Int J Syst Sci: Oper Logist 6(1):86–98
26. Chan FT, Kumar N (2007) Global supplier development considering risk factors using fuzzy extended AHP-based approach. Omega 35(4):417–431
27. Yrjölä M (2015) Uncovering executive prioritization: Evaluating customer value propositions with the pairwise comparison method. J Serv Sci Manag 8(01):1
28. Liu X, Liang M, Luo Q (2019) Innovative electric vehicle body design based on insurance institute for highway safety side impact conditions. Autom Innov. https://doi.org/10.1007/s42154-019-00070-4
29. Liu M (2006) Environmental problems in China—an overview. Chin Bus Rev 5(1):56–59
30. National Highway Traffic Safety Administration: United States Federal Motor Vehicle Standards and Regulations. https://www.safercar.gov. Accessed 9 June 2018
31. National Highway Traffic Safety Administration: Side Impact Protection. https://www.safercar.gov. Accessed 9 June 2018
32. Insurance Institute for Highway Safety: IIHS Side Impact Test Program Rating Guidelines (2006) https://www.iihs.org. Accessed 9 June 2018
33. Mall S, Anbanandam R (2022) A fuzzy analytic hierarchy process and VIKOR framework for evaluation and selection of electric vehicle charging technology for India. Transp Dev Econ. https://doi.org/10.1007/s40890-022-00150-x
34. Ogra V (2018) Use of electric vehicles to transform mass transportation in India (December)
35. Global EV Outlook 2020 (2020) Global EV Outlook 2020. https://doi.org/10.1787/d394399e-en
36. Chung SH, Kwon C (2015) Multi-period planning for electric car charging station locations: a case of Korean expressways. Eur J Oper Res 242(2):677–687. https://doi.org/10.1016/j.ejor.2014.10.029
37. Wu Y, Yang M, Zhang H, Chen K, Wang Y (2016) Optimal site selection of electric vehicle charging stations based on a cloud model and the PROMETHEE method. Energies 9(3):1–20. https://doi.org/10.3390/en9030157
38. Kwiesielewicz M (1998) A note on the fuzzy extension of Saaty's priority theory. Fuzzy Sets Syst 95(2):161–172. https://doi.org/10.1016/S0165-0114(96)00329-6
39. Chang DY (1996) Applications of the extent analysis method on fuzzy AHP. Eur J Oper Res 95(3):649–655
40. Saaty TL (2002) Decision making with the analytic hierarchy process. Scientia Iranica 9(3):215–229. https://doi.org/10.1504/ijssci.2008.017590

Chapter 6
FDI and Investment Gap in the EV Sector

6.1 Introduction by Editor

6.1.1 Introduction

The EV transition in India will require considerable investments including FDI. The chapter will document the nature and volume of investment and FDI flowing into the sector. This chapter also examines the question of barriers to investment and what can be done to make India a more attractive destination for investment.

6.1.2 Investment for EV Transition

The EV industry received a record-breaking USD 3.6 billion in funding in FY23.[1] Well-known companies such as Tata Motors, Ashok Leyland, Mahindra, etc. have set up or are setting up factories as we discuss below. The sector is witnessing the entry of foreign battery manufacturers such as Suzuki, Denso, Toshiba, Alperex, and Leclanche as well as investments by Indian companies such as Amara Raja Batteries etc. Besides, there is the emergence of a range of domestic companies and start-ups in the energy storage space. Indian EV start-ups have raised INR 4490 crores between 2014 and 19.[2] This is from over 250 businesses including startups, across all market segments.

Some notable investments in the space are noted are discussed below.

---

[1] https://sarepenergy.net/wp-content/uploads/2023/03/Investment-Landscape-of-Indian-E-Mob ility-Market-V3-FINAL.pdf.

[2] Inc42 Media, "Electric Vehicle Market Outlook Report 2020," 2020.

S. Ray and D. Mukherji (eds.), *A Primer on Electric Vehicles in India*,
https://doi.org/10.1007/978-981-97-8966-5_6

- **Tata Motors** which has the highest share in the 4-wheeler segment, began production of EVs in 2020 with most parts being imported.[3] Tata AutoComp currently produces around 50% of the components in-house.[4] It has set up a subsidiary Tata Passenger Electric Mobility Limited (TPEML) and plans to invest $1 billion in the EV business.[5] In May 2022, TPEML and Ford India signed an MoU with the Government of Gujarat for the acquisition of Ford's Gujarat-based vehicle manufacturing plant. Tata Motors raised $1 billion from TPG and Abu Dhabi state holding company ADQ for its EV unit.[6] There are 4 models that are currently sold by Tata: Nexon EV Max, Nexon EV Prime, Tiago and Tigor.[7] Tata-owned Jaguar Land Rover in partnership with Agratas will produce batteries at a large scale in Gujarat for both lithium-ion phosphate and nickel manganese cobalt chemistries.[8] Tata Chemicals Ltd has plans for large battery recycling and recovery of critical minerals from end-of-life batteries.[9]
- **Ashok Leyland**[10] has a share of 31.7% in electric buses domestically and plans to launch an electric bus in the international market later this year. It also has plans along with Switch Mobility (which is a net zero subsidiary of Ashok Leyland) for an electric light commercial vehicle (eLCV) soon.[11] Ashok Leyland and Switch Mobility have launched two small electric trucks IeV3 and IeV4 recently.[12]
- **Mahindra Group** has a share of 14.6 in the 3-wheeler segment. It has brands such as Zor Grand, and the Treo range in the EV segment.[13] Last Mile Mobility,

[3] https://www.thehindubusinessline.com/companies/tata-indias-electric-vehicle-king-takes-a-frugal-road-less-travelled/article65374865.ece.

[4] https://www.businesstoday.in/auto/story/tata-motors-frugal-plan-to-raise-ev-production-331993-2022-05-02.

[5] https://www.thehindubusinessline.com/companies/tata-indias-electric-vehicle-king-takes-a-frugal-road-less-travelled/article65374865.ece.

[6] https://www.thehindu.com/business/tata-motors-in-talks-to-sell-minority-stake-in-ev-business-report/article66543392.ece.

[7] https://www.cartrade.com/tata-cars/electric-cars/#:~:text=Some%20of%20the%20popular%20Tata,cars%20in%20the%20below%20list.

[8] https://www.livemint.com/companies/news/jlr-turns-to-tata-s-agratas-for-ev-batteries-11686593156306.html.

[9] https://www.tatachemicals.com/Asia/News-room/Press-release/tata-chemicals-lion-battery-operations#:~:text=In%20keeping%20with%20our%20entry,lithium%2Dion%20cells%2Fbatteries.

[10] https://www.livemint.com/companies/people/leyland-to-focus-capex-on-alternative-fuel-tech-11694111284078.html.

[11] https://www.cnbctv18.com/auto/ashok-leyland-auto-sales-electric-vehicle-budget-2023-focus-infrastructure-help-company-15844501.htm.

[12] https://www.thehindubusinessline.com/companies/ashok-leylands-ev-arm-switch-mobility-unveils-electric-small-trucks-in-the-scv-segment/article67281210.ece#:~:text=The%20trucks%20will%20offer%20a,this%20payload%20in%20the%20market.

[13] https://www.businesstoday.in/technology/news/story/mahindra-last-mile-mobility-becomes-indias-largest-electric-three-wheeler-maker-in-fy23-385264-2023-06-12.

its division producing many last-mile mobility solutions, merged with the parent company in 2023.[14]

- **Suzuki Corporation** formed a joint venture with Toshiba Corporation and Denso Corporation to manufacture lithium-ion batteries in India by setting up Automotive Electronics Power Pvt. Ltd (AEPPL) with a corpus of Rs. 5000 crores.[15] The production capacity of the AEPPL is around 1 GWh and is expected to produce 30 million lithium-ion cells by 2025. These batteries could be deployed in pure electric vehicles, hybrids, and electric two-wheelers. The second phase of this battery unit was commissioned recently.[16]
- **Hyundai** India's second largest car maker and biggest car exporter Hyundai Motor India Ltd has planned an investment of INR 40 billion in India to launch six electric vehicles (a mix of premium, hatchbacks, and SUVs) by 2028. Hyundai is developing a small electric car for the Indian market as a part of this plan.[17] Hyundai was one of the first overseas auto majors to launch an EV in 2019 when 15 they introduced KONA SUV to the Indian market under CKD route.
- **Mercedes-Benz** set up the first base for assembling EVs outside Germany at its plant in Chakan in Pune and it has recently launched its second electric offering EQS 580 sedan for India.[18] With this, Mercedes will also be the first and only luxury carmaker to assemble an EV in the country.
- **Aether:** A new investment of INR 4.2 billion in Ather Energy in January 2022 was announced by Hero MotoCorp.[19] Hero MotoCorp is an early investor in Ather Energy and has been a part of its growth story since 2016. Earlier, Aether Energy raised US$ 50 million led by existing backer Caladium Investment.[20]
- **Bajaj Auto** set up a facility in Pune with an investment of INR 3 billion. The company has proposed to invest more than INR 1 billion in a new unit for electric vehicles, under GoI's PLI scheme.[21] Since 2007, Bajaj Auto and KTM of Austria have been partners in the motorcycle segments and they are planning to enter the electric motorcycle segment.[22] The company is aiming to launch a new electric two-wheeler every year for the next 3–5 years. Bajaj plans to manufacture 2,50,000 units per year at its new EV-dedicated Akurdi facility in Pune.

[14] https://www.mahindra.com/news-room/press-release/en/mahindra-electric-mobility-limited-completes-merger-with-mahindra-mahindra-limited.

[15] Suzuki Motor powers EV rush in India with Rs 10,440-crore investment (Business Standard).

[16] Ibid.

[17] Hyundai to launch 6 EVs by 2028, invest INR 4000 crore towards electrification (Auto Economic Times).

[18] Mercedes Benzl India becomes the first base for Mercedes-Benz to assemble EVs outside Germany (Auto Economic Times).

[19] Hero MotoCorp lines up Rs 420 crore fresh investment in Aether Energy (Economic Times).

[20] Ather Energy raises $50 million in funding as it ramps up electric scooter production (Auto Economic Times).

[21] Bajaj Auto to invest ₹1,000 cr. under PLI scheme to accelerate EV production (Business line).

[22] Bajaj Auto, KTM plan to launch high-end electric motorcycles in India (Times of India).

- **Exide:** India's largest lead-acid battery maker, Exide Industries Ltd, has set up Exide Leclanche Energy, a JV with Swiss firm Lechlance, to manufacture lithium-ion batteries in India.[23] Exide plans a greenfield EV battery manufacturing facility near Bengaluru with an investment of INR 60 billion.[24]
- **Ola Electric** with a valuation of US$5 billion, and has raised about US$866 million with the latest being the Series C funding round of US$200 million in January 2022.[25] Ola has announced the setup of a global centre with USD 100 million, 'Ola Futurefoundry' for advanced engineering and vehicle design in Coventry, UK. The centre will work with the Ola Campus in Bengaluru in the areas of two-wheeler and four-wheeler EV design, advanced high-performance automotive engineering, digital and physical modelling, and cell technologies.[26]
- **TVS Motors** along with BMW, its partner since 2013, is investing in the future mobility space with a special focus on EVs.[27] This partnership will see the development of exclusive products by both companies on the common platform and both will retail their products globally. TVS has also invested in Bengaluru-based, high-performance electric motorcycle manufacturer, Ultraviolette Automotive in December 2021.[28]

Apart from that, several others are lined up[29]:

- Reliance Group plans to invest INR 750 billion to set up four Giga factories in India to make batteries for EVs, solar photovoltaic modules, advanced energy storage batteries, electrolysers, and finally, fuel cells in these Giga factories.
- Globally General Motors and Honda have joined hands to expand their collaboration on electric vehicles and batteries. Honda Motorcycle and Scooter India (HMSI) is planning to launch multiple electric models in India.
- Indian automotive components manufacturer Lucas TVS has signed an MoU with a US-based start-up to establish a semisolid lithium-ion cell manufacturing facility in Chennai. The first phase of the planned 10 GWh facility will start production in the second half of 2023.
- Amara Raja Batteries, one of India's largest automotive battery makers, has opened its lithium-ion cells manufacturing facility in Tirupati, Andhra Pradesh. Amara Raja Batteries invested in Slovakia-based advanced battery manufacturer, InoBat in 2021.
- Ola Electric has made a strategic investment in Israeli battery technology company, StoreDot.

---

[23] Exide-Leclanche JV to scale up production at India battery plant (Reuters).

[24] Exide announces Rs 6000 Cr investment in lithium-ion cell unit in Bengaluru (Economic Times).

[25] Ola Electric raises $200 million in fresh funding at $5 billion valuation (Economic Times).

[26] Ola to invest over $100mn to set up engineering and design centre in UK (The Week).

[27] TVS Motor and BMW Motorrad to jointly develop electric vehicles (Business Line).

[28] Ibid.

[29] BYD has invested about US$ 200 million in India and was planning further investments. The investment has been stopped. https://www.livemint.com/politics/policy/govt-rejects-chinese-byd-megha-engineerings-1-billion-proposal-to-set-up-ev-plant-11690200288132.html.

- In April 2021, Hero MotoCorp and Taiwan's Gogoro announced their plans to establish a joint venture to introduce Gogoro's industry-leading battery-swapping network in India. Hero MotoCorp invested about US$285 million in this joint venture to develop a battery-swapping platform.
- Sun Mobility, Log9, Himadri Chemicals, Tirupai Grpahite, Himadri Chemicals, and Godi India are some of the Indian companies that are working on developing advanced cell technologies for the Indian EV sector.
- The Chatterjee Group (TCG) is setting up a battery research laboratory in Kolkata at their Research Institute for Sustainable Energy, being developed by TCG Centre for Research, Education, Science, and Technology (TCG Crest).

6.1.3 Charging Infrastructure

Multiple stakeholders are exploring various business models and planning partnerships to set up EV charging.

- A subsidiary of Energy Efficiency Services Ltd (EESL), Convergence Energy Solutions Limited (CESL) is working on an initiative for setting up 810 electric vehicle charging stations along 16 expressways and National Highways covering 10,275 km across India. CESL will install 590 chargers of 50 kW capacity that will be available every 25 km and 220 chargers of higher 100 kW capacity, every 100 km.87 CESL has also set up a solar-powered charging station for electric vehicles in the Ladakh region.
- Tata Power has partnered with Hyundai Motors to install its range of fast chargers (DC 60 kW) at Hyundai's existing 34 EV dealer locations across 29 cities in India.
- Automovil, the mobility start-up in India for car servicing has partnered with Midgard Electric to set up 500 EV charging stations in 11 cities in India.
- Exicom, one of India's leading power solutions companies with a presence in the telecom, mobility, and stationary energy segment has crossed the milestone of installing over 5000 EV chargers including AC and DC chargers across 200 cities in India.
- EV charging network provider Statiq has partnered with the auto major Mahindra & Mahindra to provide EV passenger car charging solutions across the country.
- Hero Electric has announced a partnership with Charzer, a Bengaluru-based EV charging startup, to build 100,000 charging stations across India.
- UK-based L-Charge has agreed on a long-term collaboration with Mumbai-based Ziptron Gas Technology to enhance the EV charging infrastructure network in India.
- Reliance BP Mobility Limited, operating under the brand name Jio-BP, is working with multiple demand aggregators, original equipment manufacturers (OEMs), and technology partners with a vision of becoming the leading EV charging infrastructure player in India. BP bought a 49% stake in over 1400 petrol pumps and 31 aviation turbine fuel (ATF) stations owned by Reliance for US$1 billion in

2019. Jio-BP has recently entered into a partnership with Mahindra. Starting with 16 cities, Jio-BP will install DC fast chargers at Mahindra's dealership network and workshops across the country. With these chargers being open to the public, the partnership will benefit all stakeholders in the EV value chain. Jio-BP is also working with BluSmart, the EV ride hailing service provider in Delhi and the National capital region, which has a fleet of more than 500 electric cars and over 300 charging stations.

6.1.4 Barriers to Investment

Certain barriers to investment remain. Moerenhout et al. (2022) examine India's present EV ecosystem with a focus on drivers of and barriers to investment. The study is based on in-depth consultations with experts[30] on (1) the state of play of India's EV market and policies, (2) the main challenges to further improving its EV ecosystem, (3) government and investor initiatives to alleviate those challenges, and (4) additional challenges to and solutions for accelerating investment. It notes that the main barriers to investment include: an absence of clear targets for EV uptake, and insufficient fiscal incentives for charging infrastructure to upscaling investment in electric mobility in India.[31]

6.1.5 Conclusion/Development of Capability in the Sector/Gaps

An estimate by Singh et al. (2020) reveals that the investment needs of the production and charging infrastructure (till the financial year 2029–30) are over $180 billion for achieving the target laid out for EVs by India. As we discussed in the chapter, a lot of investment has been received and more is in the pipeline. The discussion of policy to attract investment and the PLI scheme will be discussed in the next chapter, Chap. 7.

Further major gaps in creating the complete EV ecosystem are seen on the supply side. The automotive industry needs a robust supply chain involving hundreds of parts, technologies, and raw materials. For EVs, we see major issues in the creation of a supply base for batteries, BMS (Battery management system), motors, controllers, and power electronics. This is mainly due to the lack of investment in these industries to create capacities. The industry needs to invest in these industries to meet the expected volumes coming up. Most of these industries are capex-heavy and yield returns over long periods of time, typically after 5–7 years. Hence it is imperative that investments are put in place now as the volumes are picking up so supplies of

[30] 59 EV experts including policymakers, investors, and companies, were consulted as well as results from an online survey with experts and companies included.

[31] About half of the respondents felt this way.

critical components are not a constraint to the Original Equipment Manufacturers (OEMs).

Machine Generated

Desai and Patel (2023) examine the deployment of EV infrastructure in Indian cities using GIS. Verma et al. (2022) examine the factors behind the adoption of electric vehicles in Bengaluru. Bhat et al. (2021) study the decision of Indian consumers in buying electric vehicles. Majumdar et al. (2014) study the electric vehicle infrastructure needed in the city of Kolkata and observe that there are economic and environmental benefits. The paper by Sheth and Sarkar (2021) computes the social benefit and costs associated with the electric bus rapid transit system in Ahmedabad. Sowmya and Sankaranarayan (2022) study the location of charging stations based on the travel time, queue time, recharging time, and the cost of battery recharging. Chu et al. (2020) examines the adoption of electric vehicles in the Korean context.

References

Ahluwalia, P. (2022) Bajaj Auto, KTM plan to launch high-end electric motorcycles. Times of India, July 27. Link: https://timesofindia.indiatimes.com/auto/electric-bikes/bajaj-auto-ktm-plans-to-launch-high-end-electric-motorcycles-in-india/articleshow/93159625.cms

Balachandar, G. (2022) Bajaj Auto to invest ₹1,000 cr under PLI scheme to accelerate EV production. Hindu Business Line, January 21. Link: https://www.thehindubusinessline.com/companies/bajaj-auto-to-invest-rs1000-crore-under-pli-scheme-to-accelerate-ev-future-growth-plans/article64928179.ece

Balachandar, G. (2023) Ashok Leyland's EV arm Switch Mobility unveils electric small trucks in the SCV segment. Hindu Business Line, September 7. Link: https://www.thehindubusinessline.com/companies/ashok-leylands-ev-arm-switch-mobility-unveils-electric-small-trucks-in-the-scv-segment/article67281210.ece#:~:text=The%20trucks%20will%20offer%20a,this%20payload%20in%20the%20market.

Balakrishnan, P. (2022) Aether Energy raises $50 million in funding as it ramps up electric scooter production. Economic times, October 19. Link: https://auto.economictimes.indiatimes.com/news/two-wheelers/scooters-mopeds/ather-energy-raises-50-million-in-funding-as-it-ramps-up-electric-scooter-production/94959144?utm_source=Mailer&utm_medium=newsletter&utm_campaign=etauto_news_2022-10-20&dt=2022-10-20&em=c3VkaXB0by5zZW5AYXN0ZXJpeC1pLmNvbQ==

Barman, A. (2021) TPG drives into Tata Motors EV business with $1 billion investment. October 13, Link: https://economictimes.indiatimes.com/industry/renewables/tpg-drives-into-tata-motors-ev-business-with-1-billion-investment/articleshow/86966881.cms

Barooah, S. B. and Bhatia, S. (2022) India becomes the first base for Mercedes Benz to assemble EVs outside Germany. Economic Times, October

1. Link: https://auto.economictimes.indiatimes.com/news/india-becomes-the-first-base-for-mercedes-benz-to-assemble-evs/94558705

Business Line (2021) Electric vehicles. Switch Mobility's $200-m outlay to power Ashok Leyland's EV plan. July 29. Link: https://www.thehindubusinessline.com/companies/switch-mobility-plans-150-200-mn-investments-to-spur-its-ev-ambitions/article35587354.ece

Business Line (2022) TVS Motor and BMW Motorrad to jointly develop electric vehicles. December 15. Link: https://www.thehindubusinessline.com/companies/tvs-motor-bmw-motorrad-to-expand-tie-up-for-joint-development-of-electric-vehicles/article37960144.ece

Car Trade (n.d) 4 Tata Electric cars in India, Link: https://www.cartrade.com/tata-cars/electric-cars/#:~:text=Some%20of%20the%20popular%20Tata,cars%20in%20the%20below%20list.

ET Auto (20,211) Ashok Leyland, IIT Madras researchers join hands to develop hybrid EVs using turbine technology, October 10, Link: https://auto.economictimes.indiatimes.com/news/commercial-vehicle/ashok-leyland-iit-madras-researchers-join-hands-to-develop-hybrid-evs-using-turbine-technology/94760213?utm_source=twitter_web&utm_medium=social&utm_campaign=socialsharebuttons

ET Auto (2021b) Hyundai to launch 6 EVs by 2028, invest INR 4000 crore towards electrification. December 8. Link: https://auto.economictimes.indiatimes.com/news/hyundai-to-launch-6-evs-by-2028-invest-rs-4000-crore-towards-electrification/88155582#:~:text=4%20min%20read-,Hyundai%20to%20launch%206%20EVs%20by%202028%2C%20invest%20INR%204000,it%20has%20in%20ICE%20vehicles.

ET Auto (2022) Mahindra Group and British International Investment commit USD 500 million for electric SUV space. Link: https://auto.economictimes.indiatimes.com/news/passenger-vehicle/uv/mahindra-group-and-british-international-investment-commit-usd-500-million-for-electric-suv-space/94415879

ET Tech (2022) Ola Electric raises $200 million in fresh funding at $5 billion valuation. January 25. Link: https://economictimes.indiatimes.com/tech/funding/ola-electric-raises-200-million-in-fresh-funding-report/articleshow/89088158.cms

Inc42 Media (2020) "Electric Vehicle Market Outlook Report 2020," 2020. Link https://inc42.com/reports/electric-vehicle-market-outlook-report-2020/

Jha, S. N. (2022) Mahindra to explore using Volkswagen's MEB component for future EVs. Financial Express. March 20. Link: https://www.financialexpress.com/business/express-mobility-mahindra-to-explore-using-volkswagens-meb-component-for-future-evs-2535028/

Kaushik, C. (2022) Bajaj EV Plans Revealed – 1 New Chetak Every Year for Next 3 -5 Years. ICN. March 29. Link: https://www.indiacarnews.com/news/bajaj-to-launch-a-new-ev-every-year-for-next-3-5-years-51490/#:%7E:text=The%20a utomaker%20plans%20to%20manufacture,spotted%20testing%20in%20the%20country

Majumdar, A. (2022) Suzuki Motor powers EV rush in India with Rs 10,440-crore investment. Business Standard. March 20. Link: https://www.business-standard.com/article/companies/maruti-suzuki-to-invest-104-bn-in-india-to-make-evs-and-batteries-122032000224_1.html

Mint (2023) Govt rejects Chinese BYD-Megha Engineering's $1 billion proposal to set up EV plant, July 25. Link: https://www.livemint.com/politics/policy/govt-rejects-chinese-byd-megha-engineerings-1-billion-proposal-to-set-up-ev-plant-11690200288132.html

Moerenhout, T., Goel, S., Aneja, D. A., Bansal, A., Ray, S. (2022) Investor Perspectives on Accelerating Growth in the EV Ecosystem. Chapter 1. In Moerenhout, T., Goldar, A., Ray, S., Shingal, A., Goel, S., Agarwal, P., Jain, S., Thakur, V., Tarun., Gaurav, K., Aneja, DA., Bansal, A. "Understanding Investment, Trade, and Battery Waste Management Linkages for a Globally Competitive EV Manufacturing Sector." Report. New Delhi. 2022. https://icrier.org/publications/understanding-investment-trade-and-battery-waste-management-linkages-for-a-globally-competitive-ev-manufacturing-sector/

Mukherjee, S. (2022) Hero MotoCorp Lines up Rs. 420 crore fresh investment in Aether Energy. January 14. Link: https://economictimes.indiatimes.com/industry/renewables/hero-motocorp-lines-up-rs-420-crore-fresh-investment-in-ather-energy/articleshow/88903088.cms

News 18. (2022) Tata Motors to Invest Rs 15,000 Cr in EV Segment in Next 5 Years, Develop 10 New Products. March 17. Link: Tata Motors is planning to invest INR 150 billion (£1.5 billion) in the EV segment in the next five years (News 18).

Panday, A. (2020) Suzuki JV to invest ₹3,715 crore in second phase of battery venture. Livemint. January 1. Link: https://www.livemint.com/companies/news/suzuki-jv-to-invest-3-715-crore-in-second-phase-of-battery-venture-11577900471545.html

Sachdev, A. (2023) Leyland to focus capex on alternative fuel tech. Livemint. September 8. Link: https://www.livemint.com/companies/people/leyland-to-focus-capex-on-alternative-fuel-tech-11694111284078.html

Shah, A. (2022) Exide-Leclanche JV to scale up production at India battery plant. Reuters. May 6. Link: https://www.reuters.com/world/india/exide-leclanche-jv-scale-up-production-india-battery-plant-2022-05-06/

Singh, V. P., Chawla, K., Jain, S. (2020) Financing India's Transition to Electric Vehicles: A USD 206 Billion Market Opportunity (FY21–FY30). New Delhi: Council on Energy, Environment and Water.

Shenoy, S., Nair, P., and D'Souza, N. (2023) Ashok Leyland bets on electric buses, eLCVs for next growth push. CNBC. February 3. Link: https://www.cnbctv18.com/

auto/ashok-leyland-auto-sales-electric-vehicle-budget-2023-focus-infrastructure-help-company-15844501.htm

SMMT (2021) Opportunities in the Indian Electric Vehicle Sector, Report. Link: https://www.smmt.co.uk/wp-content/uploads/sites/2/Opportunities-in-the-Indian-Electric-Vehicle-Sector.pdf

South Asia Regional Energy Partnership (SAREP) And Invest India (2023) Investment Landscape of Indian E-Mobility Market, March 2023.

The Economic Times (2021) TVS Motor Company makes fresh investment in Ultraviolette Automotive. December 21. Link: https://economictimes.indiatimes.com/industry/renewables/tvs-motor-company-makes-fresh-investment-in-ultraviolette-automotive/articleshow/88253492.cms

The Economic Times (2022) Exide announces Rs 6000 cr. investment in lithium-ion cell unit Bangalore. March 31. Link: https://economictimes.indiatimes.com/industry/renewables/exide-announces-investment-of-rs-6000-crore-in-karnataka/articleshow/90569891.cms

The Week (2022) Ola to invest over $100 mn to set up engineering and design centre in UK, January 27. Link: https://www.theweek.in/news/biz-tech/2022/01/27/ola-to-invest-over-100mn-to-set-up-engineering-and-design-centre-in-uk.html

6.2 Machine Generated Summaries

Disclaimer: The summaries in this chapter were generated from Springer Nature publications using extractive AI auto-summarization: An extraction-based summarizer aims to identify the most important sentences of a text using an algorithm and uses those original sentences to create the auto-summary (unlike generative AI). As the constituted sentences are machine selected, they may not fully reflect the body of the work, so we strongly advise that the original content is read and cited. The auto generated summaries were curated by the editor to meet Springer Nature publication standards. To cite this content, please refer to the original papers.

Machine generated keywords: india, government india, consumer, purchase, station, charge station, charge, perceive, electric mobility, incentive, mobility, mission, adoption evs, vehicle consider, public charge.

6.2.1 Leveraging GIS to Deploy Demand-Driven Public Charging Infrastructure in an Indian Metropolitan City [1]

This is a machine-generated summary of:

Desai, Apurvkumar; Patel, Chetan R.: Leveraging GIS to deploy demand-driven public charging infrastructure in an Indian Metropolitan city [1]

Published in: Spatial Information Research (2023).

Link to original: https://doi.org/10.1007/s41324-023-00514-4.

If you want to cite the papers, please refer to the original.

For technical reasons we could not place the page where the original quote is coming from.

Abstract-Summary

"To overcome this, it is essential to deploy public charging infrastructure in a demand-driven approach to meet the needs of EV users."

"This study provides a framework for developing a public charging infrastructure plan using geographic information systems, where demand for public charging is calculated based on ten indirect parameters such as population density, petrol stations, public spaces, and others."

"The analysis results produced a public charging demand map for 2025, which showed high EV charging demand in the core city region and along key transport corridors."

"The geospatial analysis is a useful tool for determining optimal locations for public charging infrastructure deployment, which can help decision-makers use the city's resources optimally."

Introduction

"With the help of the present study, the authors propose a framework for preparing a demand-driven public charging infrastructure plan for any city."

"The primary goal of the study is to create a demand-driven charging infrastructure deployment plan, for which factors from the literature review are to be identified as indicators of public charging demand."

"The selection for these factors has been based on criteria indicating EV population, availability of home/private charging, traffic attracting capacity, amount of time spent at these locations, etc. At the end of the literature review, population density, shopping centers, public transportation stations, petrol stations, recreational spaces, public places, parking places, points of interest, building density, and building height are identified as indicators of public charging demand."

"The proposed framework can help optimize the scarce resources of ULBs, especially in developing and underdeveloped countries, by planning the demand-driven public charging infrastructure."

Methodology

"The charging demand of the study area is required for the preparation of such plan."

"This charging demand can be calculated using the EV population in the study area, EV users' travel needs, and the specifications of EVs on the market."

"Ten factors acting as indicators of public charging demand are finalized for the present study."

"The data required for the analysis include the boundary of the study area and data related to factors identified from the literature review."

"The total charging demand of the city is then multiplied by the share calculated to arrive at the charging demand for each cell of the hexagonal grid of the study area."

"Public charging demand for Surat in 2025 is expected to be 42.4 MWh."

"The results of the geospatial analysis can be used for developing a demand-driven charging infrastructure deployment plan for the study area."

Results and Discussion

"To generate public charging demand map, the public charging demand of the study area is to be multiplied by the calculated share of charging demand."

"The public charging demand for 2025 is expected to be 42.4 MWh for the study area."

"The public charging demand is multiplied by the share calculated in the above section, providing the public charging demand map for 2025 for Surat city."

"Besides the core city, high public charging demand is also observed along the critical transport corridors of the city, like the SW-NE corridor, towards the south and east direction of the city."

"Most existing and proposed charging stations, i.e., 37 out of 50, are in the cells having the highest public charging demand (301–535 kWh/day)."

"From the remaining stations, 10 and 3 charging stations are deployed in cells with public charging demand of 201–300 kWh/day and 155–200 kWh/day, respectively."

Conclusion

"The present study aimed at developing a framework to prepare a demand-driven public charging infrastructure deployment plan, with a case study of Surat, India."

"The study identified ten factors from the literature review as indicators of public charging demand."

"The study area was divided into a hexagonal grid to derive the public charging demand for each cell."

"For determining the share of each cell in the public charging demand of the city, the geometric mean score for all cells was calculated considering equal weightage for all the factors."

"This proposed framework can also be used for developing demand-driven public charging infrastructure plans for other cities."

"The present study uses indirect factors to calculate the public charging demand in the area."

"The framework suggested in the present study can be used to develop demand-driven public charging infrastructure deployment plans for other cities."

6.2.2 Factors Influencing the Adoption of Electric Vehicles in Bengaluru [2]

This is a machine-generated summary of:

Verma, Meghna; Verma, Ashish; Khan, Mahim: Factors Influencing the Adoption of Electric Vehicles in Bengaluru [2].

Published in: Transportation in Developing Economies (2020).

Link to original: https://doi.org/10.1007/s40890-020-0100-x.

Copyright of the summarized publication:

Springer Nature Switzerland AG 2020.

If you want to cite the papers, please refer to the original.

For technical reasons we could not place the page where the original quote is coming from.

Abstract-Summary

"Multiple response analysis gave the potential barriers and benefits, which the consumer might consider before EV adoption."

"Cross-tabulation was further used to identify significant associations between various factors affecting the purchase decision of EVs."

"Perceived Environmental benefits, as well as financial incentives, are found to act as key motivating factors for EV adoption."

"The outcomes of this study will be useful in understanding consumer preferences, which will help policymakers in shaping policies related to the promotion and adoption of EVs in India."

Introduction

"Norway being the world's largest electric vehicle market has achieved high market penetration of EVs not only through wide packages of incentives like free parking, exemption from road tax, etc., but they found it necessary to reduce the cost difference between the CVs and EVs [3], as consumers will be willing to purchase this new technology only if they perceive it to be better than CVs."

"The objectives of this paper are: (1) To carry out an extensive literature survey and identify the potential factors influencing the EV purchase behavior. (2) To identify the key motivating factors and deterrents associated with the EV adoption in the Indian context. (3) To formulate policies that may drive commuters' towards EV adoption and forsake their conventional vehicles."

"A brief understanding of the literature is done to identify dominating factors that may influence the potential early adopters of EVs, thereby identifying the critical incentive and other motivating factors associated with EV adoption including the deterrents that a consumer may recognize before shifting to EVs."

Literature Review

"Afroz [4] in their study found a significant association of EV adoption with age and education, assuming that the young generation is more environmentally conscious and educated people are aware of the negative impacts of conventional vehicles and hence are more likely to adopt EVs."

"The cost and second the performance of the vehicle [5, 6] identified five factors influencing the PD of any new product, one of them being a relative advantage, which is the degree to which the new technology is perceived to be better than the traditional one [7] also claimed that the perceived advantages of using CVs over battery-driven vehicles (BDV) can act as a strong barrier and can weaken the propensity of consumers towards EV adoption."

"Although low running cost and environmental benefits are perceived to be imperative motivators of EV adoption, the high initial cost, limited charging infrastructure, and battery range, on the other hand, are considered strong deterrents of EVs [7–9]."

"In Norway, EVs have achieved substantial market penetration through economic incentives, thereby reduced the high upfront cost, exempted from road tax, parking fees, etc., and have been provided access to bus lanes [10]."

Methodology

"It mainly depends on five major factors i.e., Relative Advantage, Compatibility, Complexity, Trialability, and Observability."

"Since EVs is a pretty much new technology in India, trialability was therefore not addressed in the existing study."

"The most important attitudinal factors related to Relative advantages, compatibility, and complexity and observability-associated with EV adoption (PD) were identified in the existing literature and a comprehensive well-structured questionnaire was developed."

Questionnaire Survey and Study Hypothesis

"Conceptual framework, the questionnaire had four sets of questions as discussed below: Stresses that the rate of adoption surges with Financial incentives, by increasing relative advantages of new technology and, therefore, the second section analyzed the influence of demand incentives on the EV purchasing behavior; literature shows that the financial incentives like Govt."

"Hypothesis 2 Consumers' environmental concern has a significant association with their willingness to adopt EVs."

"Hypothesis 3 Consumers' awareness of air pollution due to CVs is significantly associated with their willingness to purchase EVs over CVs."

"Hypothesis 4 Consumers' perception of potential environmental benefits of EVs is significantly associated with their willingness to purchase EVs over CVs."

"Hypothesis 5 Consumers' perception of potential improvement in the quality of life due to EVs is significantly associated with their willingness to purchase EVs over CVs."

Survey Summary

"The sample represents a considerably higher percentage of females (72.1%) than males (27.9%)."

"60% of the sample is married with the majority, being private-sector employees (59.7%)."

"More than half of the respondents belonged from families with at least four members and nearly 26.1% and 44.2% of the respondent's family income fall under 2 lakh (students, private and self-employees) and 8 lakh (private sector employees) respectively."

Analysis and Results

"Consumers' extent of environmental consciousness was tested on their willingness to purchase EVs and a significant association was found between the two ($p = 0.48$)."

"81.1% of the respondents who perceive CVs as the main threat to air pollution are willing to buy EVs in the future."

"Respondents were also asked if the quality of life could be improved through EV adoption and was found significantly associated with willingness to adopt EV ($p = 0.006$)."

"Fisher's exact test showed a significant association between EV awareness of the respondents with their willingness to shift towards the electric vehicle in the future ($p = 0.000$)."

"The initial purchase cost of EV was found to have a significant association with the family income ($p = 0.34$), where 86.5% of the respondents belonging to income group of 2–8 lakh, showed concern towards the initial cost of EV."

Conclusion

"That environmental awareness and consciousness have a significant impact on the consumers' behavioral intention but although people seem to recognize the environmental benefits of EV adoption, they still seemed concerned for inadequate charging stations and time-consuming recharging."

"The low cost of ownership and price of EV were recognized as the least appealing and doubtful attribute, fuel efficiency, on the other hand, was ranked number one as the most appealing attribute for EV adoption."

"The relative advantages of EVs pertaining to its fuel efficiency are evidently affecting the adoption behavior of the consumers."

"A large number of the population still seems to be unaware that EV batteries can be charged at home."

"Along with Financial incentives which are important motivating factors for boosting EV market share, taxes should be increased on CVs ownership."

Future Studies

"This limitation can be addressed in future studies while also studying other associated aspects of EV adoption."

"While making a purchase decision a consumer tends to choose among different alternatives considering various parameters associated with each of the alternatives."

"O access the consumer's complex decision-making behavior, future research is required in identifying the parameters that a consumer might hold vital before choosing among various alternatives."

"Before purchasing a new vehicle a consumer may compare parameters such as purchase cost of different alternatives, fuel cost, maintenance cost, financial and non-financial incentives, safety, comfort, environmental consciousness, etc. Therefore, multiple-criteria decision-making analysis (MCDA) would be a viable method of evaluating conflicting criteria in making a purchase decision."

"Not many studies have been conducted on the effect of fuel economy on the EV market share in India."

6.2.3 Measuring and Modelling Electric Vehicle Adoption of Indian Consumers [11]

This is a machine-generated summary of:

Bhat, Furqan A.; Verma, Meghna; Verma, Ashish: Measuring and Modelling Electric Vehicle Adoption of Indian Consumers [11].

Published in: Transportation in Developing Economies (2021).

Link to original: https://doi.org/10.1007/s40890-021-00143-2.

If you want to cite the papers, please refer to the original.

For technical reasons we could not place the page where the original quote is coming from.

Abstract-Summary

"In spite of financial and non-financial incentives being provided to both consumers and manufacturers in India, the acceptance and adoption of electric vehicles remains very low."

"environmental enthusiasm, technological enthusiasm, anxiety (or perceived risk), social image, social influence, perceived benefits, performance expectancy and facilitating conditions on the consumers' intention to adopt electric vehicles."

"Results indicate that environmental enthusiasm, technological enthusiasm, social image, social influence, perceived benefits, and performance expectancy are positively related to adoption intention whereas facilitating conditions and anxiety have negative influence on a consumer's intention to adopt an electric vehicle."

"The results from this study can be useful to planners and policy makers to improve the adoption rate of electric vehicles."

Introduction

"Among all the modes of transportation, road vehicles are the primary sources of greenhouse gas emissions and other pollutants contributing to about 75% of the global carbon dioxide emissions from the transportation sector [12] and is responsible for three-quarters of the urban air pollution [13]."

"Advancements in energy-efficient sustainable mobility technologies like electric vehicles are considered as one of the essential responses to minimizing the greenhouse gas emissions from the transportation sector [14, 15]."

"It is extremely important to study the factors affecting electric vehicle adoption since the consumers are more likely to purchase a product if it meets their requirement and specifications [16]."

"Although the factors affecting the adoption of electric vehicles in developed economies have been studied, such research remains extremely scarce in the context of developing economies such as India."

"This study tries to fill this gap by evaluating the effects of various latent factors that affect consumers' behaviour towards electric vehicle adoption."

Theoretical Background and Research Hypothesis

"To studies analyzing the effect of consumers' personality and values [17–19], there are studies that have explored other factors affecting the consumers' intention to adopt electric vehicles."

"We assume: Environmental enthusiasm positively affects consumer's intention to adopt an electric vehicle."

"We assume: Technology enthusiasm positively affects consumer's intention to adopt an electric vehicle."

"Technology enthusiasm negatively effects consumers' anxiety towards adopting an electric vehicle."

"We assume: Anxiety negatively affects consumer's intention to adopt an electric vehicle."

"We assume: Social image positively affects consumer's intention to adopt an electric vehicle."

"Social influence positively affects consumer's intention to adopt an electric vehicle."

"Perceived benefits positively affect consumer's intention to adopt an electric vehicle."

"We assume: Performance expectancy positively affects consumer's intention to adopt an electric vehicle."

"We assume: Facilitating conditions positively affect consumer's intention to adopt an electric vehicle."

Data and Methodology

"This study comprises of eight perceived variables to analyze consumers' intention to adopt an electric vehicle."

"For the purpose of this study, the intention to adopt indicates the consumers' willingness to purchase an electric vehicle."

"To ensure reliability and rationality, the items employed to measure the perceived latent variables were modified from earlier studies to fit the context of this research."

"Consumers' intention to adopt an electric vehicle was evaluated by four items adapted from earlier studies [20, 21]."

"The items employed for this construct probed respondents about their interest in electric vehicles and their assessment of these vehicles."

"Items for measuring environmental and technological enthusiasm were derived from the research study of He and others [22]."

"Wordings of some of the items on the constructs were modified to fit the context of this research."

Data Analysis

"This curiosity and interest in new and developing technologies among potential consumers has been captured by the latent construct of technology enthusiasm, which might trigger the adoption of electric vehicles."

"Hypothesis 2 measured the impact of consumers' technological enthusiasm on their intention to adopt electric vehicles (H2a) and its effects on how buyers perceive the risks of buying electric vehicles (H2b)."

"The latent construct of anxiety measures the risks and fears that consumers perceive as being associated with purchasing an electric vehicle."

"Hypothesis 6 analyzed the impact of consumers' perceived benefits of purchasing and driving an electric vehicle on their adoption intention (H6a) and anxiety towards electric vehicles (H6b)."

"The latent factor of performance expectancy measures the consumers' opinion of how well the electric vehicles perform."

"Hypothesis 8 investigated the effect of facilitating conditions on consumers' anxiety (H8b) and intention to adopt electric vehicles (H8a)."

Discussions and Conclusions

"This study is one of the very few studies that examine the intention to adopt electric vehicles among the potential vehicle buyers of India and the factors that influence their decision-making process."

"The study found environmental enthusiasm, technological enthusiasm, social image, social influence, perceived benefits, performance expectancy, facilitating conditions to have a positive impact on consumers' intention to adopt electric vehicles."

"Bearing in mind that anxiety has a negative impact on the adoption intention, government and electric vehicle manufacturing companies need to mitigate the negative perceptions of electric vehicles among potential buyers."

"Schemes and policies can be made to subsidize the purchase of electric vehicles, reduction in registration charges and road taxes for electric vehicles, exemption from toll charges for electric vehicle users, free or reserved parking spaces in and around the city centroids etc. Third, the results showed a positive influence of technology enthusiasm, environmental enthusiasm, social image, and social influence on adoption intention."

6.2.4 Study on Possible Economic and Environmental Impacts of Electric Vehicle Infrastructure in Public Road Transport in Kolkata [23]

This is a machine-generated summary of:

Majumdar, Deepanjan; Majhi, Bijoy Kumar; Dutta, Atanu; Mandal, Ratan; Jash, Tushar: Study on possible economic and environmental impacts of electric vehicle infrastructure in public road transport in Kolkata [23].

Published in: Clean Technologies and Environmental Policy (2014).

Link to original: https://doi.org/10.1007/s10098-014-0868-7.

If you want to cite the papers, please refer to the original.

For technical reasons we could not place the page where the original quote is coming from.

Abstract-Summary

"Transport sector in India accounts for 20% of total commercial energy demand of the country, of which a considerable amount is consumed in the form of liquid and gaseous fuel."

"A major part of these fuels are imported by the Government."

"The present study examines the possible impacts on economy and environment by the implementation of battery electric vehicles (BEVs) along with the conventional road transport system in metropolitans with a case study of Kolkata."

"The major impact has been observed in controlling the vehicular emission with a decrease in CO_2 level by 26.27 t per day, on replacement of only 2% of the present public transport by suitable BEVs."

"In case of fuel economy, the infrastructure has shown the potential in reducing the consumption of diesel and autogas (LPG) by 11,654 and 3,256 L per day, respectively."

Introduction

"In India, among the total freight and passenger load, 60 and 80% of the loads, respectively, are shared by road transport (Ramachandra and Shwetmala [24])."

"For the reduction of subsidy burden, Government of India has recently decided to decrease the subsidy from diesel for bulk consumers, thus affecting the public road transport costs, rail fares, and even infrastructure related activity."

"If the use of conventional liquid fuel in the passenger road transport sector can be reduced or substituted by a non-conventional energy source, then considerable amount of foreign exchange can be saved making a positive impact."

"Battery-operated electric vehicles have been considered as substitute for conventional vehicles in the passenger road transport sector."

Literature Review and Transport Scenario

"In India, the total number of registered motor vehicles was 141.8 million by the end of the fiscal year 2010–2011 (Road Transport Year Book [25])."

"Unlike the personalized mode, the share of buses in total registered vehicles has declined from 11.1% in 1951 to 1.1% in 2011 (Road Transport Year Book [25])."

"The share of road vehicles in Kolkata city was about 34% of that of the total number of registered motor vehicles in the state of West Bengal (Ramachandra and Shwetmala [24])."

"By the end of March, 2009, the number of registered buses and taxis plying on the streets of Kolkata has been 6,938 and 32,826 (Road Transport Year Book [26]), respectively."

"The primary reason behind the decrease in the number of buses and taxis has been the cancelation of vehicles having registrations prior to 01.01.1993 (Road Transport Year Book [25]) as per the law enforced by the Government."

Methodology and Calculations

"The energy consumption pattern has been studied for different types of public transport vehicles in Kolkata."

"Pertaining to the present condition of passenger transport in Kolkata, the rising fuel cost and environmental concern, a nominal share of the conventional public road transportation modes has been considered for replacement by battery-operated counterparts."

"The battery ratings for the electric vehicles have to be optimized."

"The data on distance traveled and energy consumed for the present conventional vehicles have been studied."

"The range of an electric vehicle has to be determined firstly by the battery stack installed and the load applied."

"These batteries have been acceptable mostly for their use in electric vehicle application as these can be reused very efficiently by laser removal of SEI (solid electrolyte interface)."

Feasibility of Electric Vehicle Infrastructure

"The average cost of charging the battery using solar photovoltaic technology has been found to be about Rs."

"2,451.84 for energy consumption per day of an electric bus."

"12.77 per kWh as the charging cost using solar energy, energy consumption cost has been Rs."

"A 150 Ah battery has been found sufficient, and the approximate energy consumed per day has been 6 kWh (http://xgddc.en.alibaba.com/product/387013173210462555/electric_rickshaw.html?tracelog=cgsotherproduct1)."

"76.62 per day per vehicle using solar energy for charging @ Rs."

"An auto-rickshaw using LPG consumes about 6 l of autogas per day at a cost of about Rs."

Charging Strategies

"For smooth running of battery-operated electric vehicles, the required infrastructure of charging stations has to be established."

"Making the system more profitable and feasible, from both economic and technical aspect, battery-swapping technology can be implemented, thus eliminating the time consumed by the buses for charging."

"Electricity for charging the batteries has to be supplied by the charging stations."

"This process can be implemented in various ways, either by establishing new charging stations or remodeling the existing fuel stations to incorporate the charging infrastructure."

Charging Stations

"The present fuel stations have been taken as potential destinations for installation of charging infrastructure with both conventional and non-conventional sources."

"If 70% of this area is considered for installing solar panels, then the average electricity generation for each fuel station is about 585 kWh per day."

"To supply the required demand for 2% replaced vehicles, at least 71 such remodeled fuel stations or charging stations or a combination of both (with the same area as that of fuel station) will be required."

Economic Impact

"To calculate the financial savings that the infrastructure can offer, it has been assumed that 2% of the buses, taxis, and auto-rickshaws using conventional fuel (diesel) are replaced by battery-operated electric vehicles."

"Buses have an average consumption rate of 50 l of diesel a day per vehicle thus the quantity of diesel saved if one bus is replaced with an electric vehicle is 50 l. Therefore, the amount of diesel saved in a day for 2% replacement of buses is 4,250 l. Hence, there is a saving of Rs."

"The consumption of diesel can be reduced by 42, 53,710 l and LPG by 11, 88,440 l a year by replacing 2% of conventional vehicles with battery-operated electric vehicles in Kolkata."

"The benefits offered by this infrastructure can be increased even more by replacing a greater number of conventional vehicles by the battery-operated electric vehicles."

Environmental Impact

"These pollutants have been formed mainly due to the combustion of the fuel used for vehicle propulsion (in this case, diesel)."

"By replacing 2% diesel-using vehicles by battery-operated electric vehicles in Kolkata, diesel consumption can be reduced by 11,654 l per day."

"The emission of pollutants from the transport sector can be reduced further by replacing more vehicles using conventional fuel with battery-operated electric vehicles."

"Majority of the pollutants emitted by the transport sector (automobiles) contain carbon."

"If the vehicles using conventional fuels can be replaced by battery-operated electric vehicles, the carbon-containing pollutants released into the atmosphere can be reduced."

Application of BEV Infrastructure Model in Other Major Cities in India

"The proposed electric vehicle infrastructure model has the potential to check the issues of increasing petroleum-based fuel consumption and pollutants emission."

"The fuel consumption and travel pattern for the vehicles have been considered to be same as that of the case study."

"Such models, if implemented globally, may address the present issue of maintaining air quality as well as reducing the use of fossil fuels."

Conclusions

"The study has shown that the electric vehicle infrastructure may be implemented along with the conventional road transport system and it has the potential to be beneficiary from both economic and environmental aspects."

"The reduction in the amount of diesel and LPG consumption per day by replacing 2% of the present passenger road transportation mode shows the feasibility of implementing parallel electric vehicle infrastructure."

"There is a major reduction in emission of pollutants from transport sector if we can come down in favor of this alternative."

"The calculations involving some of the major metropolitans in India have clearly reviewed the importance of technological advancement required in the transport sector, to check the fuel economy of the country as well as meeting the pollution standards."

6.2.5 Social Benefit Cost Analysis of Electric Bus Transit for Ahmedabad [27]

This is a machine-generated summary of:

Sheth, Anal; Sarkar, Debasis: Social Benefit Cost Analysis of Electric Bus Transit for Ahmedabad [27].

Published in: Transportation in Developing Economies (2021).

Link to original: https://doi.org/10.1007/s40890-021-00116-5.

Copyright of the summarized publication:

If you want to cite the papers, please refer to the original.

For technical reasons we could not place the page where the original quote is coming from.

Abstract-Summary

"This paper is an attempt to compute social benefit cost analysis (SBCA) for the proposed electric bus rapid transit system (e-BRTS) in Ahmedabad."

"Electric bus rapid transit system is being envisaged for Ahmedabad and partial funding has recently been sanctioned by the Government of India for the purpose."

"This research focusses on the social benefit cost analysis of e-BRTS for the specific scale and context proposed for the city."

"The authors conclude that the electric BRTS has substantial societal benefits and may be viewed as a fruitful investment for the city."

Introduction

"As is the case with other transport infrastructure projects, the decision of city authorities to transition to electric mobility from conventionally fuelled vehicles will involve various feasibility concerns including the project's financial feasibility."

"The three wheeler segment in the country mainly targeting the auto rickshaws is a paratransit mode that helps last mile connectivity but this system of transport is unregulated and the shift to electric auto rickshaws will be largely the choice of the owners of these vehicles."

"The objective of this paper is twofold: To identify and present a framework for performing social benefit cost analysis (SBCA) of electric bus mobility To perform and present the results of SBCA for the proposed electric bus rapid transit system in Ahmedabad The paper has been organized as follows: first, literature pertaining to SBCA studies in transportation sector has been reviewed and presented."

Literature Review

"The thoroughness and comprehensiveness of benefit cost studies of transport investment projects during the time in Britain [28] were discussed and recommendations from the study included considerations of equity, use of discounted cash flow techniques and sensitivity analysis for benefit cost studies of the future."

"The methodological issues of benefit cost analysis [29] for capital intensive infrastructures were identified and a model which involves comprehensive evaluation based on six different categories of economic benefits can be applied for ex-ante project assessment was proposed."

"In a nutshell, the challenges in benefit cost analysis, the lack of socio-environmental value, the economies of scale and integration and the particular context of BCA technique for electric vehicles were discussed in various research literatures."

"Reviewing the available literature, it was found necessary that a benefit cost analysis technique that addressed the economic, social and environmental aspects of electric public transport vehicles was required."

Methodology and Conceptual Framework

"of vehicles on road (diverted traffic): O_{dv}; ceasing of diesel bus operation on road: O_{db} (c) Environmental benefit for users; due to reduction in air pollution: P_a; due to reduction in noise pollution: P_n (d) Benefits to users due to savings in travel time (e) Benefits due to reduction in accidents (B_{ra}); benefits due to reduction in major fatal accidents (B_f); benefits due to reduction in minor non-fatal accidents (B_{nf}); benefits due to reduction in vehicle damage (B_{vd}) (f) The associated costs that need to be considered for computation of the SBCAof e-BRTS are primarily. (a) Infrastructure investment costs(IC); capital cost of corridor development (C_{cd}); capital cost of charging infrastructure development (C_{ci}); (b) (c) (d) (e) (f) (g) Cost of system operation and maintenance (OM); cost of electricity (C_e); cost of major maintenance (C_{ma}); cost of minor maintenance (C_{mi}); cost of skilled manpower (drivers, maintenance staff etc.) (C_{smp}); cost of unskilled manpower (helpers) (C_{ump}) (h) Cost of additional electric power generation (Ep); cost of power plant installation (C_{pp}); cost for transmission and distribution for the additional power (C_{td}) Considering the above benefits, present worth or annual worth analysis was carried out for the net benefits or advantages to the users and agency."

Case Study and Analysis

"The socio-economic benefit cost analysis of Ahmedabad BRTS [30] has revealed an estimate for the total vehicle operating cost saving of INR 766 million for the year 2018–19 for the entire system."

"The annual BRTS corridor maintenance cost for the civil components was expected to be INR 133 million for the year 2018–19 for the then proposed 88.8 km corridor, as per study by Ahmedabad Janmarg Limited."

"of 325 kWh electric buses, the total cost of battery replacement incurred every 5 years is INR 1358 million."

"For a total of 32,548 daily vehicle km travelled by the e-BRT, this cost can be estimated at INR 701 million annually."

"Total present worth of costs = INR 22,806 million (USD 330.52 million) Social benefit cost ratio of proposed e-BRTS for 12% discount rate and 25 years life cycle is 1.45."

Results and Discussions

"The social benefit cost analysis study carried out for a life cycle of 25 years for the e-BRTS network under consideration in Ahmedabad revealed a benefit to cost ratio greater than unity indicating that it may be a positive investment for the city if the various socio economic factors are also evaluated and considered."

"The study addresses and evaluates various costs and benefits for the case of e-BRTS for Ahmedabad with special efforts to account for the benefits due to reduction in noise pollution and the costs for power generation which have been neglected by most other studies in similar domain."

"Whereas the costs associated with the implementation of the BRTS in general such as corridor development, infrastructure maintenance cost, intelligent transport management system (ITMS) equipment replacement costs and system operation and maintenance costs are significant and unavoidable, some additional costs of charging infrastructure development, bus procurement and replacement, battery replacement and electric power generation are also expected for the e-BRTS to be in place."

Conclusions

"This research presents a comprehensive SBCA framework for evaluating the implementation case for electric bus transport systems."

"The authors conclude that the case of electric BRTS implementation for the selected network in the city is favourable as the present worth of benefits exceeds those of the costs."

"The major cost (51%) involved for implementation of an e-BRT project is for infrastructure development (amounting to INR 11,549 million for the studied case), its major subcomponent being corridor development."

"A significant finding of the detailed existing and future benefit cost components for the context-specific case studied revealed that the e-BRTS the additional benefits foreseen due to e-BRTS (86% or INR 28,250 million) far outweigh the additional costs (20% or INR 4,514 million) incurred for its successful implementation."

"The future scope of research may address efforts to develop a stronger and self-sustaining financial model and business case that may enable electric bus mobility in Indian cities."

6.2.6 Optimal Scheduling of Electric Vehicle Charging at Geographically Dispersed Charging Stations with Multiple Charging Piles [31]

This is a machine-generated summary of:

R, Sowmya; Sankaranarayanan, V.: Optimal Scheduling of Electric Vehicle Charging at Geographically Dispersed Charging Stations with Multiple Charging Piles [31].

Published in: International Journal of Intelligent Transportation Systems Research (2022).

Link to original: https://doi.org/10.1007/s13177-022-00316-2.

If you want to cite the papers, please refer to the original.

For technical reasons we could not place the page where the original quote is coming from.

Abstract-Summary

"Assigning a charging station to each electric vehicle is considered as an optimization problem to minimize travel time, queue time, recharging time, and cost of energy for battery recharging."

"The objective function is constrained to the reachability of the electric vehicle with the available state of charge of the battery to the allotted charging station without violating the maximum permissible depth of discharge limit and allowable charging rate."

"The optimization model empowers the users to prioritize the function variable based on travel requirements and battery specifications in different case studies using the Opposition-Based Marine Predator Algorithm."

"The solution obtained and the analysis of results show that the proposed strategy significantly reduces travel time, queue time, recharging time, and energy cost while fulfilling the constraints imposed."

Introduction

"The solution is obtained using an algorithm that considers a single vehicle at a time, calculates waiting time and travelling distance for each charging station from various options available, and then assigns the most suitable charging station."

"While the algorithm developed in this study fulfills the primitive objective of charging station assignment, it does not consider the effect of charging stations assigned to other vehicles while calculating the waiting time."

"The work considers multiple piles with different charging rates at each CS and considers SoC levels of batteries; however, the work calculates waiting time based on the probability of arrival and departure of vehicles instead of actual data obtained through communication among vehicles and charging stations [32, 33]."

"Algorithms for calculating distances, waiting time, and obtaining an optimal solution for the problem of charging station assignment are discussed in this section."

Problem Formulation

"The monetary cost paid for recharging the vehicle battery depends on the charging rate at the pile where it is being recharged."

"Queue length at each charging pile is mainly influenced by recharging the time of individual vehicles and the arrival rate of vehicles to that particular charging pile."

"Information such as vehicle location, battery data, charge rates, and price details of charging stations can be directly obtained from real-time communication with vehicles and charging stations."

"K_{c1} and K_{c2} are the conversion factors for driving and queuing time to match the weight of charging price in the cost function."

Proposed Opposition-Based Marine Predator Algorithm (OBMPA)

"The initial MPA solution is uniformly distributed in the search space instead of the initialization steps of several other heuristic algorithms."

"Step 1: Initialize the population X randomly, with a size of N Step 2: Put on the OBL scheme and produce opposite solutions, then choose the suitable solutions Step 3: Execute the MPA to update the population position of each individual and find the optimal current position as per the best value of the fitness Step 4: Execute the next iteration in MPA."

"Solution parameters, total time, waiting time, the recharging cost are updated in the optimization algorithm, and this iterative process is repeated until the solution converges to an optimum value."

"The proposed opposition-based marine predator algorithm in which random uniform populations are generated to search optimum solutions within a specified search area is used to obtain the solution."

Simulation Results and Discussions

"In order to validate the performance of the proposed OBMPA in assigning the CS optimally to all EVs, the obtained results in terms of total time, queuing time, and charging cost are compared with the results obtained by other selected algorithms."

"As far as the total time and charging cost is concerned OBMPA performs better than MFO to achieve overall minimum value among all the other algorithms."

"Negative sign indicates that the minimized charging cost of the respective algorithm is less than OBMPA."

"In order to validate the performance of the proposed OBMPA in assigning the CS optimally to all EVs, the obtained results in terms of total time, queuing time, and charging cost are compared with the results obtained by other selected algorithms."

"Even though the primary objective is to minimize the total time, the charging cost obtained by the OBMPA is less than other selected algorithms."

Performance Comparison

"The statistical data results, such as the minimum (Min), maximum (Max), Mean, standard deviation (STD), and run-time (RT), can be used to evaluate the effectiveness of the proposed OBMPA."

"Compared to other algorithms, the recommended OBMPA can achieve the smallest maximum (Max) value in most cases."

"When comparing the OBMPA to other algorithms, it can be seen that it achieves the lowest STD values."

"Based on the generated Min value, all algorithms provide a ranking using FRT with a 95 percent significance level."

"In all of the case studies, the OBMPA is ranked first among all of the algorithms that were considered."

Conclusion

"A charging strategy is proposed using an opposition-based marine predator algorithm (OBMPA) to solve the charging pile assignment to electric vehicles."

"A performance comparison was also carried out to verify the effectiveness of the proposed strategy and the proposed OBMPA."

"The results concluded that the proposed strategy and OBMPA significantly reduce the total time spent by electric vehicles traveling to their destination."

"The trade-off between optimization of total time and cost based on the preference given by electric vehicle users during charge requests is also included in the proposed strategy."

"Studies, the proposed scheduling strategy can also be extended to a large number of EVs with more possible constraints."

"The proposed strategy may also be optimized using multi-objective and many-objective optimization algorithms to get a best-compromised solution."

6.2.7 Patience and the Adoption of Electric Vehicles: An Application of the Dual-Self Model [34]

This is a machine-generated summary of:

Chu, Wujin; Kim, Hyunsik; Im, Meeja: Patience and the adoption of electric vehicles: an application of the dual-self model [34].

Published in: Journal of Business Economics (2020).

Link to original: https://doi.org/10.1007/s11573-020-01018-5.

If you want to cite the papers, please refer to the original.

For technical reasons we could not place the page where the original quote is coming from.

Abstract-Summary

"This study examined the importance of consumer patience and car usage on adoption of electric vehicles."

"This research applied and extended the dual-self model of Fudenburg and Levine (Am Econ Rev 96(5):1449–1476, 2006) to the EV purchase context."

"The study extended the original model by relaxing the assumptions on the relation between self-control cost and net benefit from action."

"The modeling proved two results: that patient individuals and individuals with greater car usage are more likely to be early adopters of electric vehicles."

"Data collected from electric vehicle owners in Korea were compared to data from internal combustion vehicle owners."

"Two key variables, patience and car usage, were measured and regressed on the dependent variable of vehicle type owned."

Introduction

"Even though the above-mentioned variables could have greater importance as factors affecting EV adoption, the focus of this study is on the role of patience, because patience can increase net benefit from usage and encourage investment in a product that takes longer time to pay back."

"It is shown that individuals with low self-control costs are more likely to purchase an EV, which has both a longer payback period, and inconveniences associated with usage."

"We first relax an assumption in the original model which states that net benefit from action (i.e., EV adoption) is independent of self-control cost, so that in this paper, net benefit from action is increasing in patience."

"The results of the logistic regression show supporting evidence for the hypotheses that patience (i.e., lower self-control cost) and greater car usage lead to earlier EV adoption."

Dual Self Theory of Electric Vehicle Purchase

"The procrastination model is a decision problem where an agent decides whether to carry out a chore today or procrastinate."

"If a person procrastinates, the chore is delayed to the next period, and the same decision problem is repeated until the chore is carried out, at which point the decision ends."

"Decision problem, a stationary solution is derived, which means that the agent decides on a decision rule, ex ante, to do the chore if the weather is worse than a certain critical level."

"In the EV purchase context, the original procrastination model can be interpreted as follows."

"The decision is whether to purchase an EV at time t or procrastinate."

"For EVs, prices do not necessarily decrease because the purchase price is affected by government incentives."

"The expected utility of the procrastinating decision can be derived."

Empirical Study: Electric Vehicle Purchase in Korea

"The survey consisted of questions on EV ownership information, usage behavior, usage satisfaction, environmental concern, patience and demographic variables."

"The sample was collected in March 2017, just four months apart from the EV owner survey."

"In order to test Hypotheses 1 and 2, a logistic regression analysis was conducted with patience and car usage as independent variables and EV ownership as the dependent variable (Yes = 1, No = 0)."

"For the car usage variable, respondents were asked both daily miles and annual miles driven on their automobile."

"While patience is a psychological variable, car usage is an economic variable, so EV adoption is affected by both economic considerations as well as psychological ones."

"To see if there are meaningful differences in the car usage between the two groups, daily use time and annual miles driven between the two groups were compared."

"This again shows that higher car usage leads to EV adoption."

Conclusion and Discussions

"This study extended the procrastination model of Fudenburg and Levine [35], and showed that patience and greater car usage will lead to early adoption of EVs."

"It is the first study to provide empirical support for the dual-self model in the context of environmentally-friendly behavior, by analyzing data from early adopters of EVs and comparing them with data from non-adopters."

"Through modeling and empirical study, patience and car usage, both individual characteristics, are shown to be factors affecting EV adoption."

"Since the results of the study emphasizes the importance of patience in EV adoption, manufacturers and governments can look for ways to reduce the barriers to EV adoption, even for impatient individuals."

"One of the limitations of this study is that, due to tractability, we were not able to model other important variables affecting EV purchase, such as environmental concern, altruistic behavior and image signaling."

Bibliography

1. Desai A, Patel CR (2023) Leveraging GIS to deploy demand-driven public charging infrastructure in an Indian Metropolitan city. Spat Inform Res. https://doi.org/10.1007/s41324-023-00514-4
2. Verma M, Verma A, Khan M (2020) Factors influencing the adoption of electric vehicles in Bengaluru. Transp Dev Econ. https://doi.org/10.1007/s40890-020-0100-x
3. Figenbaum E, Fearnley N, Pfaffenbichler P, Hjorthol R, Kolbenstvedt M, Jellinek R, Emmerling B, Bonnema GM, Ramjerdi F, Vågane L, Iversen LM (2015) Increasing the competitiveness of e-vehicles in Europe. Eur Transp Res Rev. https://doi.org/10.1007/s12544-015-0177-1
4. Afroz R, Rahman A, Masud MM, Akhtar R, Duasa JB (2013) Malaysian perception and attitude towards electric vehicle. In: 2nd International conference on mechanical, automotive and aerospace engineering, p 5
5. Potoglou D, Kanaroglou PS (2007) Household demand and willingness to pay for clean vehicles. Transp Res Part D Transp Environ 12:264–274. https://doi.org/10.1016/j.trd.2007.03.001
6. Rogers EM (2004) Diffusion of Innovations. Everett M Rogers Am Anthropol 65:1146–1147. https://doi.org/10.1525/aa.1963.65.5.02a00230
7. Bockarjova M, Steg L (2014) Can protection motivation theory predict pro-environmental behavior? Explaining the adoption of electric vehicles in the Netherlands. Glob Environ Change 28:276–288. https://doi.org/10.1016/j.gloenvcha.2014.06.010
8. Carley S, Krause RM, Lane BW, Graham JD (2013) Intent to purchase a plug-in electric vehicle: a survey of early impressions in large US cites. Transp Res Part D Transp Environ 18:39–45. https://doi.org/10.1016/j.trd.2012.09.007

9. Lebeau K, Van Mierlo J, Lebeau P, Mairesse O, Macharis C (2013) Consumer attitudes towards battery electric vehicles: a large-scale survey. Int J Electr Hybrid Veh 5(1):28–41. https://doi.org/10.1504/ijehv.2013.053466
10. Figenbaum E (2017) Perspectives on Norway's supercharged electric vehicle policy. Environ Innov Soc Transit 25:14–34. https://doi.org/10.1016/j.eist.2016.11.002
11. Bhat FA, Verma M, Verma A (2021) Measuring and modelling electric vehicle adoption of Indian consumers. Transp Dev Econ. https://doi.org/10.1007/s40890-021-00143-2
12. IEA (2020) Tracking Transport 2020. IEA, Paris
13. Lefevre B, Enriquez A (2014) Transport sector key to closing the world's emissions gap. World Resources Institute. https://www.wri.org/insights/transport-sector-key-closing-worlds-emissions-gap. Accessed 6 May 2021
14. Bakker S, Jacob Trip J (2013) Policy options to support the adoption of electric vehicles in the urban environment. Transp Res Part D Transp Environ 25:18–23. https://doi.org/10.1016/j.trd.2013.07.005
15. Sierzchula W (2014) Factors influencing fleet manager adoption of electric vehicles. Transp Res Part D Transp Environ 31:126–134. https://doi.org/10.1016/j.trd.2014.05.022
16. Sheth JN, Newman BI, Gross BL (1991) Why we buy what we buy: a theory of consumption values. J Bus Res 22:159–170. https://doi.org/10.1016/0148-2963(91)90050-8
17. Bunce L, Harris M, Burgess M (2014) Charge up then charge out? Drivers' perceptions and experiences of electric vehicles in the UK. Transp Res Part A Policy Pract 59:278–287. https://doi.org/10.1016/j.tra.2013.12.001
18. Glerum A, Stankovikj L, Thémans M, Bierlaire M (2014) Forecasting the demand for electric vehicles: accounting for attitudes and perceptions. Transp Sci 48:483–499. https://doi.org/10.1287/trsc.2013.0487
19. She ZY, Sun Q, Ma JJ, Xie BC (2017) What are the barriers to widespread adoption of battery electric vehicles? A survey of public perception in Tianjin China. Transp Policy 56(March):29–40. https://doi.org/10.1016/j.tranpol.2017.03.001
20. Ajzen I, Fishbein M (1980) Understanding attitudes and predicting social behavior. Prentice-Hall, Englewood Cliffs
21. Barbarossa C, Beckmann S (2015) A self-identity based model of electric car adoption intention: a cross-cultural comparative study. Elsevier, Amsterdam
22. He X, Zhan W, Hu Y (2018) Consumer purchase intention of electric vehicles in China: the roles of perception and personality. J Clean Prod 204:1060–1069. https://doi.org/10.1016/j.jclepro.2018.08.260
23. Majumdar D, Majhi BK, Dutta A, Mandal R, Jash T (2014) Study on possible economic and environmental impacts of electric vehicle infrastructure in public road transport in Kolkata. Clean Technol Environ Policy. https://doi.org/10.1007/s10098-014-0868-7
24. Ramachandra TV, Shwetmala (2009) Emissions from India's transport sector: statewise synthesis. Atmos Environ 43:5510–5517
25. Road Transport Year Book (2012) Transport research wing. Ministry of Road Transport & Highways, Government of India
26. Road Transport Year Book I (2011) Government of India
27. Sheth A, Sarkar D (2021) Social benefit cost analysis of electric bus transit for Ahmedabad. Transp Dev Econ. https://doi.org/10.1007/s40890-021-00116-5
28. Barrell DWF, Hills PJ (1972) The application of cost-benefit analysis to transport investment projects in Britain. Transportation 1:29–54
29. Florio M, Sirtori E (2016) Social benefits and costs of large scale research infrastructures. Technol Forecast Soc Change 112:65–78
30. Socio economic costs benefit analysis of Ahmedabad BRTS, Ahmedabad Janmarg Limited (2016) http://shodhganga.inflibnet.ac.in/bitstream/10603/189686/14/15_chapter6.pdf. Accessed 03 Jul 2018
31. Sowmya R, Sankaranarayanan V (2022) Optimal scheduling of electric vehicle charging at geographically dispersed charging stations with multiple charging piles. Int J Intell Transp Syst Res. https://doi.org/10.1007/s13177-022-00316-2

32. Sowmya R, Sankaranarayanan V (2019) An optimal model for electric vehicle battery charging and discharging scheduling strategy. In: 2019 national power electronics conference, NPEC 2019. https://doi.org/10.1109/NPEC47332.2019.9034691
33. Kadam VS, Sowmya R, Sankaranarayanan V (2019) Optimal coordinated charging strategy for electric vehicles at geographically distributed charging stations. In: 2019 National Power Electronics Conference, NPEC 2019. https://doi.org/10.1109/NPEC47332.2019.9034766
34. Chu W, Kim H, Im M (2020) Patience and the adoption of electric vehicles: an application of the dual-self model. J Bus Econ. https://doi.org/10.1007/s11573-020-01018-5
35. Fudenburg D, Levine DK (2006) A dual-self model of impulse control. Am Econ Rev 96(5):1449–1476

Part IV
Ecosystem

Part IV will discuss the evolution of the ecosystem in India. It discusses the conditions that lead to the development of an ecosystem and how an ecosystem evolves thereafter. It also discusses the role of industrial and trade policy in this context. These issues are discussed in Chapter 7 on policies and regulations in the EV space. Chapter 8 is on financing the EV transition in India. Finally, Chapter 9 discusses the skills gap and employment in the EV industry.

Chapter 7
Policies and Regulations at the Central Level and States

7.1 Introduction by Editor

7.1.1 Introduction

At the centre, there are policies like Faster Adoption and Manufacturing of (Hybrid &) Electric Vehicles in India Scheme (FAME) I and FAME II. Production Linked Incentive Scheme (PLI) scheme for Advanced Cell Chemistry has also been announced. In addition, most states have an EV policy. Some of the issues addressed in this chapter are whether these policies and regulations are conducive and enough to usher in an EV transition. The chapter will also look at gaps in policy that must be overcome.

7.1.2 Policies at the Centre

The policies at the Centre [1] include the National Mission for Electric Mobility (NMEM) [2], Faster Adoption and Manufacturing of (Hybrid &) Electric Vehicles in India (FAME India) Scheme (FAME) I [3, 4] in 2015 [5–7] and FAME II in 2019 [8, 9], and the Production Linked Incentive Scheme (PLI) [10, 11]. These policies have been analyzed in detail in Ray et al., (2022a).

Apart from these schemes, there has been support for the creation of charging infrastructure. It has been proposed that charging stations will be established on both sides of the road at an interval of about 25 km each on highways [12]. There has also been a reduction of Goods and Services Tax (GST) on EVs from a rate of 12% to that of 5%. Ministry of Power has also allowed the sale of electricity as a 'service' for charging electric vehicles. Battery-Operated Transport Vehicles and Transport Vehicles running on Ethanol and Methanol fuels were also granted exemption from the requirements of the permit. The Finance Minister announced an

S. Ray and D. Mukherji (eds.), *A Primer on Electric Vehicles in India*,
https://doi.org/10.1007/978-981-97-8966-5_7

additional income tax deduction of Rs.1.5 lakh on the interest paid on loans taken to purchase electric vehicles in the budget of 2019–20 [13, 14].

7.1.3 Policies at the State

In addition to the central government's policy initiatives, state governments have been proactively contributing to the EV policy landscape by formulating their own policies. Almost all states have released their EV policies [15, 16]. Each state's EV policy caters to different aspects of the EV sector and contains incentives and measures for different EV sector actors. The provisions contained in the states' EV policies can be classified under the following broad categories (Ray et al., 2022a):

- Supply-side incentives and measures
- Demand-side incentives and measures
- Labour force-related incentives and measures
- Research and development-related incentives and measures
- Charging stations and other network infrastructure-related incentives and measures [17]

Ray et al. (2022a) also discusses the EV policies of three Indian states namely Karnataka, Maharashtra, and Tamil Nadu [18]. Karnataka, Maharashtra, and Tamil Nadu are three of the top five states driving EV registrations.

7.1.4 Are the Policies Conducive to Developing the EV Ecosystem?

In any new industry with disruptive technology, there is a typical life cycle observed. The innovators and early adopters are in the early stage of technology. Then come the early majority with their own developed technology fit for a particular market segment. Post this the market sees the entry of majority players. This is the time when a few risk-taking companies take the lead with their innovative products. In the context of the Indian industry, we see the inflexion point reaching now, around 5 years after we have seen the electrification drive started with low-end Chinese 2 and 3-wheelers in the market. The industry scales up in the next 3–5 years and we will have high technology and quality products from several OEMs as has been discussed in Chap. 6. IEA (2022) examines the policies adopted by countries for greater EV adoption. It notes that support is needed particularly in emerging and developing countries for creating EV infrastructure. Ray et al. (2022c) suggest that linkages between industrial and investment policies should be amalgamated for greater EV adoption in India [19].

Favourable government initiatives like FAME II, the two Production Linked incentive Schemes (Auto and ACC), the budget exemptions for customs duty on capital goods machinery for manufacture, and the duty cut on lithium-ion batteries from 21 to 13% have all contributed to the sector's expansion and are expected to lower prices and further boost investment in the industry. The sector has also seen significant investment from a wide range of investors, including PE and VC funds, sector-focused funds, and international strategic assets. Components sector specific incentives especially for the Small, Micro and Medium (SMEs) could be a great idea to encourage the industry to set up EV components manufacturing in India. Since MSMEs contribute a large portion of the automotive manufacturing sector, it would start large-scale manufacturing of EV components in India.

There is a sustained push towards a green energy ecosystem with government incentives like FAME II, low registration fees, state-level incentives, vehicle scrappage policy, and PLI schemes are major boosts to the EV transition. However, are these enough to develop the ecosystem and capabilities? That remains to be seen.

Machine Generated

Das and Bhat (2022) analyse the factors behind EV adoption in other countries to draw lessons for India. The factors include a strong policy framework. Kumar and Yadav (2017) examine the critical challenges confronting e-mobility in India. Donkyu et al. (2022) examine the cost of ownership of batteries: they find that battery wear depends on travel distance in BEVs while in PHEVs, the frequency of acceleration and braking. Saqib et al. (2017) analyse the requirements of a smart grid for EV charging and identify three parameters such as tariff, nearest charging spot and congestion which can determine the choice. Using the concept of genetic algorithms Efthymiou et al. (2017) provide an approach for locating charging stations in Thessaloniki.

References

Baruah, R. (2023) Govt cuts subsidy to electric 2 wheelers. Livemint May 21. Link: https://www.livemint.com/companies/news/govt-cuts-subsidy-on-electric-2wheelers-11684690738134.html

EV Reporter (2022) PLI Scheme for ACC battery and storage: challenge and opportunities. May 2. Link: https://evreporter.com/pli-scheme-for-acc-battery-storage-challenges-and-opportunities%EF%BF%BC/

Government of Uttar Pradesh (2022) Uttar Pradesh Clean Mobility and Vehicle Manufacturing Policy. Link: https://cleanmobilityshift.com/wp-content/uploads/2022/10/New-UP-EV-Policy-2022-V10.pdf

IEA (2022) Global EV Outlook 2022. Securing supplies for an electric future.

IESA (n.d.) Government extends FAME India scheme for third time, Industry news. Link: https://indiaesa.info/news-menu/28-industry-news-management/1177-government-extends-fame-india-scheme-for-third-time

Invest India (n.d.) Production Linked Incentive https://www.investindia.gov.in/production-linked-incentives-schemes-india

Kanuri, C., Rao., R., and Mulukutla, P. (2021) A Review of Government Policies for Electric Mobility. WRI India Ross Centre for Sustainable Cities Report. Link: https://www.wricitiesindia.org/sites/default/files/Full_report_EV_State_Policy.pdf

Ministry of Heavy Industries (2019a) Mobility Solutions get a boost: National Mission on Transformative Mobility and Battery Storage approved by Cabinet. PIB. 7 March. Link: https://pib.gov.in/PressReleaseIframePage.aspx?PRID=1567807

Ministry of Heavy Industries (2019b) FAME India Scheme, PIB. 9 July. Link: https://pib.gov.in/PressReleasePage.aspx?PRID=1577880.

Ministry of Heavy Industries (2019c) Sanction for 5595 electric buses under FAME Phase-II. PIB. 8 August. Link: https://pib.gov.in/Pressreleaseshare.aspx?PRID=1581525

Ministry of Heavy Industries (2020) Promotion of Electric Vehicles. March 2. PIB. Link: https://pib.gov.in/PressReleasePage.aspx?PRID=1604991

Ray, S., Thakur, V., and Singh, S. (2022a) Green Industrial Policy: An Appraisal of Three States. Chapter 3. In Moerenhout, T., Goldar, A., Ray, S., Shingal, A., Goel, S., Agarwal, P., Jain, S., Thakur, V., Tarun., Gaurav, K., Aneja, DA., Bansal, A. "Understanding Investment, Trade, and Battery Waste Management Linkages for a Globally Competitive EV Manufacturing Sector." Report. New Delhi. 2022. https://icrier.org/publications/understanding-investment-trade-and-battery-waste-management-linkages-for-a-globally-competitive-ev-manufacturing-sector/

Ray, S., Thakur, V., and Singh, S. (2022b) Powering up electric vehicle adoption in India. Blog. March 31. Link: https://saonray.com/powering-up-electric-vehicle-adoption-in-india/

Ray, S., Thakur, V., and Singh, S. (2022c) EV Value Chain Needs To Harness Industrial, Investment Policy Linkages. Mobility Outlook, June 9. Link: https://www.mobilityoutlook.com/commentary/ev-value-chain-needs-to-harness-industrial-investment-policy-linkages/?utm_source=LinkedIn&utm_medium=Post&utm_campaign=MOSMPost9June2022&utm_content=AutoB2BEVValueChainNeedsToHarnessIndustrialInvestmentPolicyLinkages

Footnotes/Endnotes

[1] The Automotive Mission Plan 2006–16 (AMP 06–16) laid down a 10-year roadmap for the automotive industry covering every aspect of its growth. The approach and ethos of AMP have been adopted even more rigorously in the development of the NEMMP 2020 plan. As has been discussed by Ray et al. (2022a), the NEMMP 2020 presented a roadmap for achieving significant penetration of efficient and environmentally friendly electric vehicle technologies in India by 2020.

[2] https://pib.gov.in/PressReleaseIframePage.aspx?PRID=1567807

[3] FAME has been extended for the third time. https://indiaesa.info/news-menu/28-industry-news-management/1177-government-extends-fame-india-scheme-for-third-time.

[4] https://pib.gov.in/PressReleasePage.aspx?PRID=1577880. The objective was to promote the manufacturing of electric and hybrid vehicle technology and ensure sustainable growth of the same. The first phase of this scheme commenced on 1st April 2015 up to April 2017 but was subsequently extended up to 31st March 2019.

[5] The incentives range from Rs.13,000 to Rs.1.38 lakh in the case of 4-wheelers, Rs.17,000 to Rs.1.87 lakh for light commercial vehicles, and, from Rs.34 lakh to Rs.66 lakh for buses. Recently, the government has cut subsidies to 2-wheelers. https://www.livemint.com/companies/news/govt-cuts-subsidy-on-electric-2wheelers-11684690738134.html

[6] In the 1st phase of the scheme, about 2.8 lakh electric vehicles were supported with total disbursement incentives of Rs.359 Crores.

[7] Phase I of the FAME India Scheme is expected to bring a multitude of advantages, including an estimated 50 million litres of total expected fuel saved due to electrification; 52,700 L of fuel saving per day; reduction of about 1.3 Lakh Kg of CO_2 per day; and total CO_2 reduction of approximately 129 million Kg.

[8] The second phase of the FAME India Scheme was implemented from April 1st, 2019 for a period of 3 years and an expected completion by March 31st, 2022. In this phase, 5095 electric buses to 64 Cities / State Transport Corporations for intra-city operation, 400 electric buses for intercity operation, and 100 electric buses for last-mile connectivity to Delhi Metro Rail Corporation (DMRC) were sanctioned. This involves Government incentives of around Rs.2800 crore. https://pib.gov.in/Pressreleaseshare.aspx?PRID=1581525

[9] https://pib.gov.in/Pressreleaseshare.aspx?PRID=1581525. It is expected to save cumulatively about 1.2 billion liters of fuel over the contract period, and avoid 2.6 million tons of CO_2 emission.

[10] https://www.investindia.gov.in/production-linked-incentives-schemes-india

[11] https://evreporter.com/pli-scheme-for-acc-battery-storage-challenges-and-opportunities%EF%BF%BC/

[12] The Department of Heavy Industries has sanctioned 2,636 EV charging stations amounting to approximately Rs.500 crore across 24 States/UTs under FAME India Scheme phase II, in 62 cities including metros, other million-plus cities, smart cities, and cities of Hilly states across the country so that there will be availability of at least one charging station in a grid of 3 km × 3 km.

[13] https://pib.gov.in/PressReleasePage.aspx?PRID=1604991

[14] Other announcements like the battery swapping scheme were made in the budget 2022–23. Ray et al. (2022b).

[15] Kanuri et al. (2021).

[16] https://cleanmobilityshift.com/wp-content/uploads/2022/10/New-UP-EV-Policy-2022-V10.pdf

[17] Ray et al. (2022c).

[18] Chosen on the basis of 1) Share in aggregate electric vehicle registrations 2) Presence of automotive clusters and automotive component clusters.

[19] https://www.hindustantimes.com/opinion/ev-adoption-guidelines-must-be-more-practical-101690638215941.html

7.2 Machine Generated Summaries

Disclaimer: The summaries in this chapter were generated from Springer Nature publications using extractive AI auto-summarization: An extraction-based summarizer aims to identify the most important sentences of a text using an algorithm and uses those original sentences to create the auto-summary (unlike generative AI). As the constituted sentences are machine selected, they may not fully reflect the body of the work, so we strongly advise that the original content is read and cited. The auto generated summaries were curated by the editor to meet Springer Nature publication standards. To cite this content, please refer to the original papers.

Machine generated keywords: india, city, order, emobility, conventional vehicle, auto, deploy, degradation, roadmap, oxide, bevs, exhaust, smart, charger, mass adoption.

7.2.1 Global Electric Vehicle Adoption: Implementation and Policy Implications for India [1]

This is a machine-generated summary of:

Das, Pabitra Kumar; Bhat, Mohammad Younus: Global electric vehicle adoption: implementation and policy implications for India [1].

Published in: Environmental Science and Pollution Research (2022).

Link to original: https://doi.org/10.1007/s11356-021-18211-w.

If you want to cite the papers, please refer to the original.

For technical reasons we could not place the page where the original quote is coming from.

Abstract-Summary

"Present transport system of conventional vehicle in India has faced challenges due to enormous amount of air pollution, health hazards to human, rising oil price, insufficient indigenous fossil fuel reserve, heavy expenditure on oil import, energy insecurity, etc. Electrical vehicle (EV) is considered to be alternatives of conventional vehicles that can overcome these shortcomings."

"The aim of the study is to get an overview of the electric vehicle policies of government of India and its state governments to find out their relevance and impact on EV adoption in India."

"Government of India has framed policies such as "NEMMP 2020," "FAME-I," "FAME-II," and Vehicle Scrappage policy."

"Seventeen of its state governments have framed EV policy."

"This will have strong impact on EV demand generation, conversion of conventional vehicles to e-vehicle resulting E-mobility transformation and EV hub in the region that is at par with other EV-developed countries in the world."

Introduction

"The immense increase in the number of motor vehicles in the modern urban transport system has posed significant threats to the air quality of India's cities."

"Given the current growth of the Indian transport sector, it is clear that India is becoming increasingly dependent on imported crude oil supply."

"Urban cities face health risks due to exhausts from vehicles, climate change and global warming due to greenhouse gas (GHG), energy security and reliance on imported oil, limited fossil fuel reserves stored in the world, and heavy spending of Forex reserves on oil imports."

"Electric vehicles (EV) can be an alternative to conventional transport in India in reducing emissions from vehicles and improving energy security which will lead to a reduction in oil imports and cost savings."

The Current Present EV Scenario

"By 2020, 67% of new electric vehicle registrations are electric vehicle batteries (BEVs)."

"China has the largest fleet (4.5 million), and Europe has the highest number of electric vehicles per year to reach 3.2 million marks."

"The estimated number of new electric vehicles registered in 2020 is 3 million."

"China took second place with 1.2 million registrations, and 295,000 new electric vehicles were registered in the USA."

"Registration for new electric vehicles in Germany and France is 395,000 and 185,000, respectively."

EV Scenario in India

"Commercial and residential complexes will be facilitated with an EV charging system (Bihar electric vehicle policy [2])."

"Dedicated charging infrastructure, development of the independent player model, and provision of charging facilities on commercial premises are encouraged (Himachal Pradesh electric vehicle policy [3]) Karnataka (approved Sep 2017): One hundred percent e-transportability in auto-rickshaws, taxi aggregators, corporate transports, and school bus/vans implementation by 2030."

"One hundred twelve charging stations for EVs will be installed in Bangalore city (Karnataka electric vehicle and energy storage policy [4])."

"One hundred percent exemption from electricity tariffs by 2025 for EV-related and charging infrastructure production units (Tamil Nadu electric vehicle policy [5])."

"Production of lithium batteries for high mileage EVs will be encouraged with the policy framework and financial incentives (Telangana electric vehicle policy [6])."

"Special concession for EV battery manufacturing, development, and charging and swapping infrastructure (Uttarakhand electric vehicle policy [7])."

Discussion

"The required basic infrastructure of manufacturing and sales of electric vehicles is already present in India."

"Growing fuel price and insufficient domestic production of crude oil is a major issue in India that will compel Indian people to avoid using conventional vehicles and attract electric vehicles."

"The major barrier of the electric vehicle observed by researchers in the Indian context is the price of the vehicle, charging infrastructure such as charger density, charging time, range, and battery cost."

"The growth and firm expansion plan in renewable energy production in India will reduce power costs and increase availability that will further increase electric vehicle usage in India."

"It means a great number of vehicles will be scrapped in India every year."

"Additional demand in two- and three-wheeled vehicle sectors will go to electric version two and three wheelers due to purchase price competitiveness and low operation cost."

Conclusion and Policy Implications

"Major factors behind EV success in these countries are found to be a strong policy framework, financial and non-financial incentives, effective charging infrastructure, electricity availability with a proper mix, green lifestyle, consumer information, and awareness campaign and wide availability of EV model, etc. The volume of the world's electric car has crossed over ten million units by 2020."

"The government of India and the majority of its state governments have taken strong initiatives in terms of strategy, policy formulation, infrastructure development, incentives, fund allocation, research and development, production, and sales at par with other EV-developed countries in the world."

"Very soon, India will dominate the EV sector like conventional vehicle production, sales, and export business provided the plan and strategies initiated by central government and state governments are implemented and executed in reality."

"Our energy security will be endangered unless found out sufficient mines for battery materials in our country or indigenously produced materials are used for EV battery production."

7.2.2 Promoting e-mobility in India: Challenges, Framework, and Future Roadmap [8]

This is a machine-generated summary of:

Shalender, Kumar; Yadav, Rajesh Kumar: Promoting e-mobility in India: challenges, framework, and future roadmap [8].

Published in: Environment, Development and Sustainability (2017).

Link to original: https://doi.org/10.1007/s10668-017-0006-x.

If you want to cite the papers, please refer to the original.

For technical reasons we could not place the page where the original quote is coming from.

Abstract-Summary

"The purpose of the research is to identify the critical challenges that are impeding the adoption of e-mobility in India."

"The study successfully identifies four different categories of challenges and proposes integrative framework for e-mobility."

"The research is novel in terms of presenting a holistic viewpoint on e-mobility in India."

"Its originality lies in identifying the major inhibitors obstructing EVs adoption in India and then suggesting the roadmap how to overcome these impediments for mass adoption of e-mobility."

Introduction

"Economic importance of the industry also originates from its linkages with host of related industries, i.e. from steel, aluminium, and rubber to component and parts manufacturers, and from suppliers and distributors to maintenance, mobility, and insurance sectors."

"It comes as little surprise that growing economies including India are putting in all the efforts to make auto industry act as catalyst to stimulate their economic growth (UNIDO [9])."

"The spurt in the financially sound middle class has also contributed to the growth of auto industry."

"In India too, the flourishing middle class has catapulted the status of auto industry from nascent to major sector which contributed 7.1% to overall GDP in 2016."

"Further, Indian auto sector accounts for 26% of industry and 49% of manufacturing GDP in the last financial year (SIAM [10])."

"The purpose of this research is to identify the challenges that e-mobility adoption faces in India."

Transition from Conventional to e-Mobility

"Up, the aspect of EVs manufacturing in India faces two major challenges: battery performance and drive technology."

"Electric battery is at heart of EVs and needs improvement in several areas (Bundesregierung [11])."

"Another contentious issue is the long charging time that batteries require to get charged which needs to be reduced for making EVs practical."

"One can charge the EVs at home or workplace, and still there is absolutely a need for a strong public charging infrastructure to help people overcome the range anxiety, the fear of being stranded in the midway owing to discharge of the battery (Egbue and Long [12])."

"In the world over today, strategies are being conceived to solve the contentious issues like generation, storage, and supply of renewable energy though access to renewable electricity for EV charging in India still remains a distant dream."

Evolution of Indian Automobile Industry and EVs

"The scenario changed in 1980s when government allowed liberalisation of the policies to allow foreign manufacturers to set up their business in India (Narayanan [13])."

"The only prominent name that produces electric cars in India is Mahindra Reva."

"The real impetus to sales of EVs came when Government of India announced national mission for hybrid and electric vehicles in 2011."

"A separate council named National council for electric mobility was set under the aegis of Ministry of Heavy Industry and Public Enterprises."

"Further at Ministry secretary level, a national board for electric mobility was constituted."

e-Mobility in India: Identification of Critical Challenges

"This method involves administration of three to four rounds of survey to the same expert panel to build the consensus on the key issues (Robinson [14])."

"The identification of critical challenges impeding the promotion of e-mobility in India is carried out in the following three phases: In the first phase of the Delphi survey, experts are asked to list various challenges and barriers that are obstructing propagation of e-mobility in India."

"The third and final phase of survey required expert to rate the challenges emerged from second phase of scrutiny according to their critical importance."

"A 2-week time frame was given to the experts in order to complete the survey and additional 3 days were provided at the end of 2 weeks."

"While the majority of challenges/issues were either accepted or modified by the experts in phase II, all three challenges under miscellaneous category were rejected by majority of experts."

Integrated e-Mobility Framework

"This lays down the foundation of our integrated e-mobility framework wherein we propose policy, innovation, and awareness as foundational components, and perception change, technology upgradation, and effective interface as operational components."

"Coming together, these components will provide the necessary fillip for a mass adoption of e-mobility in India."

"Policy, innovation, and awareness undergird our e-mobility framework."

"By providing purpose, a sense of direction, and articulating clearly the requirements, both at strategic and tactical level, these foundational components will create a groundswell support for e-mobility adoption."

"The responsibility of e-mobility operationalisation rests on the shoulders of three enablers—perception change, technology upgradation, and effective interface and integration of electricity generated from renewable resources."

"With integrated framework in place, the next section of research proposes the roadmap to help mass adoption of e-mobility within the broad contours of policy, technology, and innovation."

Roadmap for e-Mobility Adoption

"We posit that promoting an e-mobility plan in India should be designed in a manner that ensures EVs emergence as self-sustaining system."

"Government can shape the opinion and help in this regard by developing new standards/scales that help to calculate the total cost of ownership of the vehicles rather than depicting only the retail price."

"Incorporating clean energy charging system in spatial planning and urban development In order to make EVs as the default option, it is required that clean energy charging stations (electricity generated from the renewable resources) must be integrated into the spatial planning and new urban development projects, in addition to increasing their concentration in existing infrastructure."

"This will result in more regular interaction of common people with charging points and help in building the opinion that EVs can be used as a default option."

Discussion and Conclusion

"The concept of e-mobility in India is still in the nascent stage though deteriorating ecosystem and rising emission levels have put this at centre stage of debate."

"India's largest car manufacturer, Maruti-Suzuki India Limited has already started rolling out cars featuring mild hybrid technology which although does not come exactly under EVs, still nobody can deny its significance in realising of e-mobility scenario."

"This will help to spur the investment in all crucial areas of e-mobility and in time help to ensure the ultimate aim of mass manufacturing of EVs."

"E-mobility in India faces many challenges; however, its key drivers outweigh these hurdles significantly."

"Promoting innovations, building charging infrastructure, and spreading awareness about the significant role that e-mobility can play in ensuring a safe and healthy future can further catalyse the mass adoption of e-vehicle in India."

7.2.3 *A Cost of Ownership Analysis of Batteries in All-Electric and Plug-In Hybrid Vehicles [15]*

This is a machine-generated summary of:

Baek, Donkyu; Bocca, Alberto; Macii, Alberto: A cost of ownership analysis of batteries in all-electric and plug-in hybrid vehicles [15].

Published in: Energy, Ecology and Environment (2022).

Link to original: https://doi.org/10.1007/s40974-022-00256-3.

If you want to cite the papers, please refer to the original.

For technical reasons we could not place the page where the original quote is coming from.

Abstract-Summary

"In order to analyze battery degradation, various tests were utilized for both a full-battery electric vehicle (BEV) and a plug-in hybrid electric vehicle (PHEV)."

"The results demonstrate greater wear of the PHEV battery when the vehicle is under charge-sustaining mode, that is, when using the combustion engine, while driving with frequent starts and stops."

"The degradation costs of the BEV battery are generally close to optimal in every scenario, in which the main parameter affecting battery wear is average daily mileage."

Introduction

"Car makers may take advantage of the lowered cost of battery cells to increase the total capacity of battery packs by adding a greater number of cells, in order to increase the maximum mileage or driving range of their BEVs before charging."

"This paper describes a cost analysis model that includes the degradation trend of battery life over time for PHEVs and BEVs, in which the size and actual use of their batteries differ."

"The novelty of this paper is summarized as follows: Proposal of a battery cost analysis model Analysis of the battery usage of BEV and PHEV depending on mobility characteristics Analysis of the optimal usage cost of batteries in order to maximize the benefit–cost ratio Analysis of the battery degradation based on battery usage, warranty and capacity."

Background

"For light-duty vehicles, such as passenger cars, the energy size of the battery packs in EVs generally differs greatly depending on the type of vehicle: (1) usually less than one kilowatt-hour for MHEVs, (2) up to about 20 kWh for FHEVs and PHEVs, and (3) in general several tens of kilowatt-hours for BEVs, even up to or greater than 100 kWh (Blomgren [16])."

"Electrical energy and fuel consumption for these vehicle types were analyzed in Wang et al. [17], resulting in some interesting conclusions regarding the optimal size of the battery packs from an energy perspective, especially in the case of PHEVs."

"In Peterson et al. [18], the authors state that vehicle-to-grid (V2G) service increases capacity fade more quickly and, therefore, this auxiliary service should generally be avoided in order not to further aggravate the degradation of batteries due to the normal operation of BEVs."

"The cost of depreciation of a battery pack over time is commonly included in maintenance costs (Bösch et al. [19])."

Cost Model

"The simulation of the overall energy flow of such vehicles is carried out by considering the vehicle powertrain model, drivetrain model including power transmission system, and battery SOC estimator."

"ADVISOR includes detailed model coefficients of engines, electric traction motors, controllers, converters, energy storage systems, shapes of chassis, etc. We carefully scaled and tuned the specification and efficiency of the components in order to simulate the following vehicles."

"We selected Tesla Model 3 and Toyota Prius for the simulation because they were the best-selling BEV and PHEV in the United States in 2019, respectively (U.S. Plug-in Electric Vehicle Sales by Model [20])."

"This car includes an 8.8 kWh lithium-ion battery pack, which we assume of lithium-ion cells of the same type used in Model 3, in order to coherently compare the battery SOC and battery aging of the selected PHEV and BEV under the same battery characteristics."

Results

"An additional daily driving time of about 15.7 min, at those test conditions, is required to achieve the optimal daily cost from a warranty perspective for the BEV under test."

"Similar to the UDDS test, for the battery usage cost analysis we assumed two driving tests for each daily driving cycle."

"The battery usage for the US06 test is higher than the dashed line because the battery consumes energy with acceleration, deceleration and high-speed driving on highway."

"For the PHEV, the battery usage cost is lower than optimal because there are fewer acceleration and deceleration phases during highway driving than during city driving."

"Due to the short driving time, the degradation of PHEV and BEV batteries is less than optimal."

"The usage cost of a BEV battery pack is close to optimal in all driving conditions."

Conclusion

"For an optimal cost of ownership of batteries in BEVs and PHEVs, this work presented a usage cost analysis of these batteries."

"The degradation of a BEV battery is generally less than 50% of that of a PHEV battery in city traffic, but it is slightly greater in highway driving."

"Driving a PHEV with the use of a traditional combustion engine is strongly discouraged on routes with frequent starts and stops, not only for the harmful effects on environment, but also for the greater degradation of the battery."

"Battery wear in BEVs is generally more correlated to travel distance than other driving characteristics, whereas in PHEVs it largely depends on the frequency of high-speed variations involving acceleration and braking phases."

7.2.4 Smart Electric Vehicle Charging Through Cloud Monitoring and Management [21]

This is a machine-generated summary of:

Saqib, Mohd.; Hussain, Md. Muzakkir; Alam, Mohammad Saad; Beg, M. M. Sufyan; Sawant, Amol: Smart Electric Vehicle Charging Through Cloud Monitoring and Management [21].

Published in: Technology and Economics of Smart Grids and Sustainable Energy (2017).

Link to original: https://doi.org/10.1007/s40866-017-0035-4.

If you want to cite the papers, please refer to the original.

For technical reasons we could not place the page where the original quote is coming from.

Abstract-Summary

"Smart charging system of electric vehicle using cloud based monitoring and management is demonstrated in this work."

"xEVs (electric plugin hybrid, battery electric vehicles) Charging Management System is crucial for the dynamic demands of charging infrastructure, namely perspectives from automakers, electricity providers, vehicle owners and charging service providers."

"The system not only provide an execution framework for the xEVs users but also provide an optimal energy trading solution to all entities involved in a smart charging infrastructure such as charging station, aggregators, smart grid etc. The work also explains the cloud enabled bidding strategies that look for day-ahead and term-ahead markets."

Introduction

"It is the call of the hour to develop a centralized EV/PHEV charging management system which controls demand and supply problem within the dedicated time frames."

"Request from vehicle or users are processed in [22] and [23] to devise hierarchical aggregator-based hybrid vehicles charging control mechanisms that may lower the variability of demand and distributed power generation through smart charging and discharging in real time."

"Contemporary developments in the intelligent transportation systems (ITS) aided with utilities for information and communication technologies (ICT) have enabled bidirectional data collection and transport mechanics where the real-time vehicle data can be utilized for modeling fleet charging management prototypes [24]."

"Motivated by the facets of cloud computing paradigms and the need for an efficient frame work for coordinated charging of xEVs, this work proposes centralized cloud based remote xEV charging management system."

"Proposed a cloud based remote xEV charging management framework recommending optimal charging solution to the involved xEV customers."

Business Model

"An instance of data center is developed using SQL Server 2008 purposed to manage the computation and analytics tasks such as login (authentication), searching, updating information regarding energy attributes and bidding analysis."

"The cloud analytics undergoing at the data center employees computationally perceptive be data analytics to recommend smart charging options to the corresponding vehicles customer."

"The application acts not only as a recommendation engine for the xEVs Users but also provide an optimal charging solution to all entities involved in a smart charging infrastructure such as charging station, aggregators, smart grid etc. Charging stations manage their user through this application performing task such as updating charging level, places to stay, update list of selected consumer after bidding etc. This application reduce load on smart grid by solving the problem of demand and supply problem."

Software Description of the Proposed System

"There are three main entities involved in the model:- 1) Admin: The role of "Admin" will perform by any centralized unit/Power System Operator (PSO) to control the available as well as new constructed charging station and power plants."

"This entity can perform the tasks such as: a) Control over self-database such as update price for consumer, update demand for main suppliers etc. b) Management for the xEVs such as give notification when EV charged, give availability etc. c) Fix the lowest price for bidding and final the price to buy from main supplier and sell to the consumer."

"Power Plants benefits operator is the main supplier of energy who has various tasks after login such as a) View data of CSs such as consumers, average expenditure, location etc. b) Management of self-records such as update energy price, availability of energy etc. c) View and sell energy after Bidding."

Proposed System Output Parameters

"The framework may also be prone to the cyber threats thus causing in cyber-physical threats for the overall smart grid ecosystem."

"The cloud aware technologies play fundamental role underpinning varying smart grid functionalities, to automate remote management tasks such as xEV fleet management, perform event driven processes etc. Such intensive use of cyber-physical infrastructure poses serious threats with respect to privacy, authentication, intrusion and novelty of such framework [25]."

"Any breach in such a delicate system from cyber fabric may have serious consequences such as power interrupt, chaos in the fleet due to mismanagement, data breach, accidents etc. Therefore, a holistic and secure cyber framework need to be deployed to address the possible vulnerabilities presented in varying modes of execution of the application."

Case Studies

"Three case studies are presented here namely Electric Vehicle Managed Charging Demand Response, Current demand for electric vehicles, Aggregate Demand Supply System According to [26], V2G has been tested in a small number of pilots, a number of technical and regulatory issues need to be resolved before it can be widely and effectively used."

"As of February 2017, more than 580,000 EVs were sold in the United States,2 representing approximately one terawatt-hour (TWh) of annual consumption.3 According to Bloomberg New Energy Finance (BNEF), EV electricity consumption is projected to increase to approximately 33 TWh annually by 2025, and 551 TWh by 2040.4 Given the projected growth in EVs, and the increasing need for flexible grid resources, more utilities are evaluating the opportunity for managed charging."

"The National Electric Mobility Mission Plan (NEMMP) target, of 5–7 million EVs by 2020, corresponds to penetration of EVs of 15% in 2W and 19% in passenger car markets."

"For the largest segment, two wheelers, hybrid/EV penetration was estimated to be 1–2 of total 2W sales."

Conclusion

"The proposed work presents a cloud aware smart charge management framework for intelligently charging the xEV fleet."

"The application provides stakeholder specific interfaces and modules to enable an uninterrupted power flow across the whole smart grid infrastructure and simultaneously circumventing the demand peaks created by the xEV penetration."

"When installed at the xEV customer end, the cloud monitored software will be able to track the geo-distribution of optimal charging station that makes the multi-objectives met concurrently."

"Among the list of recommending parameters, the application will specific focus on the three requirements in decreasing order or priority namely minimum charging tariff, nearest charging spot and the station having minimum xEV congestion."

7.2.5 *Electric Vehicles Charging Infrastructure Location: A Genetic Algorithm Approach [27]*

This is a machine-generated summary of:

Efthymiou, Dimitrios; Chrysostomou, Katerina; Morfoulaki, Maria; Aifantopoulou, Georgia: Electric vehicles charging infrastructure location: a genetic algorithm approach [27].

Published in: European Transport Research Review (2017).

Link to original: https://doi.org/10.1007/s12544-017-0239-7.

If you want to cite the papers, please refer to the original.

For technical reasons we could not place the page where the original quote is coming from.

Abstract-Summary

"In order to prepare themselves to welcome the increased number of electric vehicles circulating in the city networks in the near future, they are expected to deploy networks of public electric vehicle chargers."

"The Electric Vehicle Charging Infrastructure Location Problem is an optimization problem that can be approached by linear programming, multi-objective optimization and genetic algorithms."

"The algorithm and a user-friendly tool have been developed in R and tested for the city of Thessaloniki."

"The results indicate that 15 stations would be required to cover 80% of the estimated electric vehicles charging demand in 2020 in the city of Thessaloniki and their optimal locations to install them are identified."

"The approach can be used to allocate charging stations at high-level, i.e. to zones, and the authors encourage its use by local authorities of other cities too, in Greece and worldwide, in order to deploy a plan for installing adequate charging infrastructure to cover future electric vehicles charging demand and reduce the electric vehicle "driver anxiety" (i.e. the driver's concern of running out of battery) encouraging the widespread adoption of electromobility."

Introduction

"Considering the importance of reducing Europe's dependence on imported oil and carbon dioxide emissions from road transport by 60% until 2020, the White Paper on Transport in 2011 [28] has set the target for the reduction of conventionally fuelled vehicles in cities."

"In order to encourage the spread of EVs in urban road transport and decrease the transition time from conventional to EVs, it is essential to eliminate obstacles and encourage policies and actions that promote the wide use of non-conventional vehicles."

"Even though, GHGs are produced for EVs construction and operation, during the production of the necessary electricity, EVs are much friendlier to the environment and their carbon footprint is at least 40% of conventional vehicles."

"The study is focused on EVs for which the charging infrastructure is essential; however, the proposed methodology and tool could be used for any type of plug-in vehicles (e.g. plug-in-hybrid vehicles)."

Literature Review

"In the area of electromobility as well, there is research interest related to the EVs charging facilities location problem, and a number of optimization processes have been proposed and developed."

"Optimization algorithms that are used to address the EV infrastructure location problem are usually based on maximization of the covered demand or minimization of the travel costs."

"Worley and others [29] developed an integer-programming model for the simultaneous solution of the vehicle routing and charging station problem, with the objective to minimize the total cost (travel, recharging and implementation)."

"They found that GAs require more time to generate a solution than other optimization algorithms, without this being a limitation for strategic decisions such are the facility location problems; moreover, the results are in many cases superior (except in the fixed charge location problem)."

"The applicability of GAs to the electric vehicle charging facility location problem has been explored in a number of research papers."

EV Penetration in Greece

"The rate of EVs penetration in Greece is low due to the following factors [30]: 1) the purchase price of an EV is higher than the price of a conventional; 2) the vehicle purchase potential of Greeks depends on imported vehicles; 3) there are no EVs of the most popular vehicle type categories of the market (economy and compact cars) available in the Greek market; 4) there are not EVs charging facilities in the Greek territory; 5) the consumers postpone the purchase of a new vehicle due to the uncertain economic situation; 6) the consumers are reluctant to buy an EV considering that the technologies are still immature."

"The Hellenic Ministry of Environment, Energy and Climate Change [31] has estimated that the penetration of EVs by 2020 will be between 2 and 7% depending on the scenario (optimistic or realistic)."

Methodology

"For the purpose of this research, we assume that the EVs with destination to one of the 124 zones of the Municipality of Thessaloniki will need to get charged if they have travelled more than a certain number of km from their origin, and thus have consumed most of their stored energy."

"Prior to the initiation the optimization algorithm, the user should specify the desired number of stations (0 – total number of locations in the input file) for the examined scenario, the average cover distance radius per station, the charging speed and the average generalized cost per km."

"The initial population was composed by randomly generated chromosomes with sum of bits equal to the number of stations, in order to restrict the search area of the algorithm and increase the optimization speed."

Conclusions and Future Prospects

"The objective of this research is to propose a solution for optimal EV charging infrastructure deployment, based on Genetic Algorithms and OD data."

"The estimation of charging demand is based on assumptions due to the scarce availability of EV data in Greece."

"The proposed optimization algorithm has been implemented in a user-friendly tool developed in R and is freely available."

"Since low EV penetration rates are more or less the case in all EU countries (as well as the whole world) this approach as well as the tool could be used to allocate charging stations at high-level in other cities too, in Greece and worldwide, in order to deploy a plan for installing adequate charging."

Bibliography

1. Das PK, Bhat MY (2022) Global electric vehicle adoption: implementation and policy implications for India. Environ Sci Pollut Res. https://doi.org/10.1007/s11356-021-18211-w
2. Bihar electric vehicle policy (2019) (draft). http://www.investbihar.co.in/Download/Draft_for_e_vechile.pdf
3. Himachal Pradesh electric vehicle policy (2019). http://himachalpr.gov.in/OnePressRelease.aspx?Language=1&ID=14445

4. Karnataka electric vehicle and energy storage policy (2017). https://kum.karnataka.gov.in/KUM/PDFS/KEVESPPolicyInsidepagesfinal.pdf
5. Tamil Nadu electric vehicle policy (2019). https://powermin.gov.in/sites/default/files/uploads/EV/Tamilnadu.pdf
6. Telangana electric vehicle policy (2017) (draft). https://www.transportpolicy.net/wp-content/uploads/2019/10/Telangana_EV_Policy_20170927.pdf
7. Uttarakhand electric vehicle policy (2019). https://evreporter.com/wp-content/uploads/2020/07/electric1576487934.pdf
8. Shalender K, Yadav RK (2017) Promoting e-mobility in India: challenges, framework, and future roadmap. Environ Dev Sustain. https://doi.org/10.1007/s10668-017-0006-x
9. UNIDO (2003) The global automotive industry value chain: what prospects for upgrading by developing countries. Sectoral Studies Series, United Nations Industrial Development Organization. http://www.unido.org/fileadmin/media/documents/pdf/Services_Modules/Automotive_Industry.pdf. Accessed 18 Mar 2016
10. SIAM (2016) http://www.siamindia.com/economicafairs.aspx?mpgid=16&pgid1=22&pgidtrail=23 . Accessed 20 Mar 2018
11. Bundesregierung (2009) German federal government's national electromobility development plan. Berlin. http://www.emic-bg.org/files/German_Federal_Governments_National_Electromobility_Development_Plan.pdf. Accessed 14 Jul 2016
12. Egbue O, Long S (2012) Barriers to widespread adoption of electric vehicles: an analysis of consumer attitudes and perceptions. Energy Policy 48:717–729
13. Narayanan K (1997) Technology acquisition, deregulation and competitiveness: a study of Indian automobile industry. Discussion Paper Series # 9703, United Nations University, INTECH Institute for New Technologies
14. Robinson JBL (1991) Delphi methodology for economic impact assessment. J Trans Eng 117(3):335–349
15. Baek D, Bocca A, Macii A (2022) A cost of ownership analysis of batteries in all-electric and plug-in hybrid vehicles. Energy Ecol Environ. https://doi.org/10.1007/s40974-022-00256-3
16. Blomgren GE (2017) The development and future of lithium ion batteries. J Electrochem Soc 164:A5019–A5025
17. Wang H, Zhang X, Ouyang M (2015) Energy consumption of electric vehicles based on real-world driving patterns: a case study of Beijing. Appl Energy 157:710–719
18. Peterson SB, Apt J, Whitacre J (2010) Lithium-ion battery cell degradation resulting from realistic vehicle and vehicle-to-grid utilization. J Power Sources 195(8):2385–2392
19. Bösch PM, Becker F, Becker H, Axhausen KW (2018) Cost-based analysis of autonomous mobility services. Transp Policy 64:76–91
20. U.S. Plug-in Electric Vehicle Sales by Model (2020) Alternative Fuels Data Center, U.S. Department of Energy. https://afdc.energy.gov/data/10567
21. Saqib M, Hussain MM, Alam MS, Beg MMS, Sawant A (2017) Smart electric vehicle charging through cloud monitoring and management. In: Technology and economics of smart grids and sustainable energy. https://doi.org/10.1007/s40866-017-0035-4
22. Alonso M, Amaris H, Germain JG, Galan JM (2014) Optimal charging scheduling of electric vehicles in smart grids by Heuristic Algorithms. Energies 7(4):2449–2475. https://doi.org/10.3390/en7042449
23. Awaad MH, Krauss H, Schmatz HD (1978) Advanced praise for the unified modeling language reference manual, 2nd edn, vol 240, no 3
24. Alam MS, Vehicle to Cloud (V2C) remote management of Electric, Hybrid and plug-in Hybrid Electric Vehicle Charging.pdf. Patent Application # 34590/DEL/2015
25. Sharif H, A survey on cyber security for smart grid a survey on cyber security for smart grid communications, no. August 2014
26. THEC for and Charging M (2017) Utilities and electric vehicles the case for. no. April
27. Efthymiou D, Chrysostomou K, Morfoulaki M, Aifantopoulou G (2017) Electric vehicles charging infrastructure location: a genetic algorithm approach. Eur Transp Res Rev. https://doi.org/10.1007/s12544-017-0239-7

28. European Commission (2011) White paper—roadmap to a single European transport area. In: Towards a competitive and resource efficient transport system, Brussels
29. Worley O, Klabjan D, Sweda ATM (2012) Simultaneous vehicle routing and charging station siting for commercial electric vehicles. IEVC
30. Spentzas K (2012) Electric vehicles in Greece: potential and prospects. Available online: http://www.heliev.gr/library_articles/K.Spentzas_environment_and_ecological_transports.pdf. in Greek
31. Hellenic Ministry of Environment, Energy and Climate Change (2012) Investigating ways of development and penetration of EVs in Greece, Athens

Chapter 8
Financing EV Transition in India

8.1 Introduction by Editor

8.1.1 Introduction

Financing the EV transition will be key. This chapter will examine various ways this financing is being done and novel ways in which capital is being raised. We have discussed whether the environment is conducive to FDI and if private domestic investment is flowing into the sector in Chap. 6.

8.1.2 Financing EV Transition

The financing requirement for the transition will be huge. It has been estimated that the cumulative capital cost of the country's EV transition will be INR 19.7 lakh crore (USD 266 billion) between 2020 and 2030 (NITI Aayog, 2021) [1]. This estimate covers vehicles, electric vehicle supply equipment (EVSE), and batteries (including replacements). The estimated size of the annual EV finance market will be INR 3.7 lakh crore (USD 50 billion) in 2030.

The report by NITI Aayog on 'Mobilising Finance for EVs in India' discusses several ways in which India can finance its EV transition. Some of these have been discussed elsewhere e.g. including EVs in priority sector lending, interest subvention (this has been discussed in Chap. 7 in the state EV policy), ensuring product guarantees by Original Equipment Manufacturers (OEMs), and ways to distribute the risks.

S. Ray and D. Mukherji (eds.), *A Primer on Electric Vehicles in India*,
https://doi.org/10.1007/978-981-97-8966-5_8

8.1.3 Financing Challenges

The challenges faced by India in financing the transition are discussed below. The challenge arises due to the absence of the manufacturing value chain as discussed in Chap. 4 [2]. Moreover, the lack of raw materials and technology poses a problem. With an increasingly unstable geopolitical climate, companies are looking to diversify their EV supply chains, and India is increasingly being perceived as a frontrunner in this regard. This is primarily due to the policies and incentive schemes launched by the Government of India and the availability of low-cost labour. The large market size for EVs, as well as India's position as a market leader in the overall automobile sector, makes the country well-positioned to become a global leader and hub for EV investment.

India's transition is different from other countries, partly due to the predominance of 2 and 3-wheelers (as opposed to 4-wheelers elsewhere) and the variability among states in EV adoption [3]. CEEW-CEF in a study, examines the barriers to the flow of finance in the sector. Among the OEMs and component manufacturers, the smaller OEMs and Tier 2 and Tier 3 suppliers face significant barriers in raising capital. Financing charging infrastructure will be the key to the EV transition. However, as argued by the CEEW-CEF report, the business case for public charging stations is the weakest since there is uncertain demand, high investment requirements, and lower cash flow. At present, the utilization rates for these charging stations are also small. This makes the case even more challenging. There are also barriers to the flow of capital for end consumers, which affects the EV adoption rates. The various state EV policies and central policies have tried to address this issue as we discussed in Chap. 7. Several solutions have also been suggested e.g. financing windows or a partial credit guarantee scheme for OEMs, or a charging infrastructure investment facility. These are specifically addressing the problem. Another way in which this can be done is through collaborations [4] and innovative solutions.

8.1.4 Collaborations

Worldwide, there are many interesting examples of collaboration [5]. In light of China's announcement of the ban on the export of certain critical minerals, the EU is working on a plan to provide electric buses to Latin America in lieu of lithium supplies [6]. In Chap. 6, we saw the instances of several firms setting up factories in India. Tesla is looking to set up a factory in India [7, 8]. Also discussed earlier, there are issues with charging infrastructure and energy storage. A venture by Reliance New Energy Limited and Faradion plan for a proposed fully integrated energy storage factory at the Dhirubhai Ambani Green Energy Giga Complex in Gujarat could be a start. Several international electric vehicle and battery projects have been initiated. These are in China, the UK, the United States, Germany, and India (UNCTAD, 2023).

8.1.5 *Innovative Solutions*

It has been estimated that India requires Rs. 162.5 lakh crores for meeting the nationally determined contributions [9]. One way this can be done is through innovative financial solutions including fintech solutions. A similar approach has to be taken in the case of EVs. One example is the pilot scheme undertaken by SIDBI called 'Mission 50K-EV4ECO' which has a direct and an indirect lending component [10]. Under direct lending, loans will be given to MSMEs while in the indirect part, NBFCs will be targeted to reduce their costs of lending. Examples from other countries can also be useful.

8.1.6 *Way Forward*

The biggest challenge the nascent EV industry faces today in India is financing this massive change. Financing is required at two levels—capital investment (availability of capital and cost of capital) and vehicle financing to the end user. In India, almost 90% of commercial vehicles (3-wheelers, trucks) are sold through financing. The financing market in India is pretty mature with 74% financing penetration and below 1% NPAs for ICE vehicles (EY, 2017). The finance portfolio is estimated at around Rs. 2600 billion. SMEV has argued for the imposition of a green tax on ICE vehicles on account of the subsidy withdrawal [11].

The way forward for EV would be to emulate the success of ICE vehicles financing model. The RBI was considering bringing EVs in the prime lending sector to enable banks and NBFCs to finance the vehicles. SIDBI has tied up with Niti Aayog, the World Bank, and Bank of Korea, and the Korean Economic Development Cooperation Fund in a new project called 'Evolve' under which vehicle financing as well as project financing would be made available to MSMEs. These projects would include fleet operators, aggregators, and vehicle financing. Initiatives like this would create a vibrant economic ecosystem for EVs in India.

Machine Generated

Some other issues have been examined e.g. the willingness to pay for a plug-in hybrid electric vehicle in the context of Kolkata by Bera and Maitra (2021). Siddiqui et al. (2016) compare the cost of battery storage systems and grid integration versus microgrids. Ghosh et al. (2022) examine the feasibility of retrofitting existing fuel stations with EV charging infrastructure in the context of the UK. Krehic (2022), examines if the use of EVs affects urban toll ring prices in the case of Norway. Kannan and Tumuluru (2022) analyze an energy trading system and find the proposed model outperformed the traditional model despite variations in the number of active EV owners and their random charging and discharging preferences.

References

E&Y (2017) Building a strategic and profitable auto finance portfolio in India, Report.

Finerty, A. (2022) Financing critical minerals on the road to a more sustainable auto sector, Standard Chartered https://www.sc.com/en/feature/financing-critical-minerals-sustainable-auto-sector/

Mint (2023) Govt rejects Chinese BYD-Megha Engineering's $1 billion proposal to set up EV plant, July 25. Link: https://www.livemint.com/politics/policy/govt-rejects-chinese-byd-megha-engineerings-1-billion-proposal-to-set-up-ev-plant-11690200288132.html

Mobility Outlook Bureau (2023) SMEV bats for green tax on ICE two-wheelers. June 6. Link: https://www.mobilityoutlook.com/news/smev-bats-for-green-tax-on-ice-twowheelers/#:~:text=SMEV%20(Society%20of%20Manufacturers%20of,pollution%2Dcreating%20crude%20oil%20imports.

Mordor Intelligence (n.d.) India Electric Vehicle Financing Market Size & Share Analysis - Growth Trends & Forecasts (2023 - 2028). Link: https://www.mordorintelligence.com/industry-reports/india-electric-vehicle-financing-market

Morenhout, T., Kohli, P. C., Goel, S., Ray, S., Janardhanan, N., and Olsen, S. H. (2023) Securing Critical Minerals Supply Chains for the Clean Energy Transition. May 2023, T20 Policy Brief. Task Force 4 Refuelling Growth: Clean Energy and Green Transitions.
https://www.orfonline.org/research/securing-critical-minerals-supply-chains-for-the-clean-energy-transition/

NDTV (2023) Tesla In Talks To Set Up Car Factory In India, EVs To Start At ₹ 20 Lakh: Report. https://www.ndtv.com/india-news/tesla-in-talks-to-set-up-car-factory-in-india-evs-to-start-at-rs-20-lakh-report-4202706#pfrom=home-ndtv_topscroll

Truitt, J.S. (2022) Apple supplier Foxconn partners with Saudi wealth fund to build EVs. Nikkei Asia, November 4. Link: https://asia.nikkei.com/Business/Automobiles/Apple-supplier-Foxconn-partners-with-Saudi-wealth-fund-to-build-EVs#:~:text=Apple%20supplier%20Foxconn%20partners%20with%20Saudi%20wealth%20fund%20to%20build%20EVs,-Ceer%20will%20license&text=NEW%20YORK%20%2D%2D%20Taiwanese%20tech,to%20diversify%20the%20kingdom's%20economy.

NITI Aayog (2021) 'Mobilising Finance for EVs in India' Link: (https://e-amrit.niti.gov.in/assets/admin/dist/img/new-fronend-img/report-pdf/mobilising_finance_for_evs_in_india_compressed-1-10.pdf).

Singh, G. (2023) Transitioning to Net Zero, the India Fintech story. Economic Times, June 4. Link: https://bfsi.economictimes.indiatimes.com/news/fintech/transitioning-to-net-zero-the-india-fintech-story/100711740

Singh, S. C., and Shah, A. (2023) Exclusive: BYD proposes $1 billion India plan to build EVs, batteries, Reuters, July 14. Link: https://www.reuters.com/business/autos-transportation/byd-seeks-nod-1-bln-plan-build-evs-batteries-india-sources-2023-07-14/.

Singh, V. P., Chawla, K., Jain, S. (2020) Financing India's Transition to Electric Vehicles: A USD 206 Billion Market Opportunity (FY21 - FY30). New Delhi: Council on Energy, Environment and Water.

The Economic Times (2023) SIDBI launches new financing solution for electric vehicle space, April 14, Link: https://economictimes.indiatimes.com/industry/renewables/sidbi-launches-new-financing-solution-for-electric-vehicle-space/articleshow/99498974.cms?from=mdr

UNCTAD (2023) World Investment Report: Investing in Sustainable Energy for All.

Valero, J. (2023) EU Working on E-Bus for Lithium Deal With Latin American Nations. Bloomberg, July 14. Link: https://www.bloomberg.com/news/articles/2023-07-14/eu-working-on-e-bus-for-lithium-deal-with-latin-american-nations#xj4y7vzkg

WEO Forum (2019) How can India transition to electric vehicles? Here's a roadmap. October 3. Link: https://www.weforum.org/agenda/2019/10/how-can-india-transition-to-electric-vehicles-heres-a-roadmap/

Footnotes/Endnotes

[1] (https://e-amrit.niti.gov.in/assets/admin/dist/img/new-fronend-img/report-pdf/mobilising_finance_for_evs_in_india_compressed-1-10.pdf).

[2] Financing critical minerals will also be important. https://www.sc.com/en/feature/financing-critical-minerals-sustainable-auto-sector/

[3] https://www.weforum.org/agenda/2019/10/how-can-india-transition-to-electric-vehicles-heres-a-roadmap/

[4] https://www.orfonline.org/research/securing-critical-minerals-supply-chains-for-the-clean-energy-transition/

[5] For example, Foxconn has partnered with the Saudi sovereign wealth fund to build EVs for Saudi Arabia. Truitt (2022).

[6] https://www.bloomberg.com/news/articles/2023-07-14/eu-working-on-e-bus-for-lithium-deal-with-latin-american-nations#xj4y7vzkg

[7] https://www.ndtv.com/india-news/tesla-in-talks-to-set-up-car-factory-in-india-evs-to-start-at-rs-20-lakh-report-4202706#pfrom=home-ndtv_topscroll

[8] Other examples could include BYD but have been stopped: https://www.reuters.com/business/autos-transportation/byd-seeks-nod-1-bln-plan-build-evs-batteries-india-sources-2023-07-14/. https://www.livemint.com/politics/policy/govt-rejects-

chinese-byd-megha-engineerings-1-billion-proposal-to-set-up-ev-plant-11690200288132.html

[9] https://bfsi.economictimes.indiatimes.com/news/fintech/transitioning-to-net-zero-the-india-fintech-story/100711740#:~:text=India%20requires%20about%20162.5%20lakh,achieving%20India's%20net%20zero%20targets.

[10] https://economictimes.indiatimes.com/industry/renewables/sidbi-launches-new-financing-solution-for-electric-vehicle-space/articleshow/99498974.cms?from=mdr

[11] https://www.mobilityoutlook.com/news/smev-bats-for-green-tax-on-ice-two wheelers/#:~:text=SMEV%20(Society%20of%20Manufacturers%20of,pollution%2Dcreating%20crude%20oil%20imports.

8.2 Machine Generated Summaries

Disclaimer: The summaries in this chapter were generated from Springer Nature publications using extractive AI auto-summarization: An extraction-based summarizer aims to identify the most important sentences of a text using an algorithm and uses those original sentences to create the auto-summary (unlike generative AI). As the constituted sentences are machine selected, they may not fully reflect the body of the work, so we strongly advise that the original content is read and cited. The auto generated summaries were curated by the editor to meet Springer Nature publication standards. To cite this content, please refer to the original papers.

Machine generated keywords: regulation service, owner, regulation, technoeconomic, operate, urban, exist, charge, realtime, revenue, smart, participate, raise, exemption, phev.

8.2.1 Commuters' Willingness-to-Pay for the Attributes of Plug-in Hybrid Electric Vehicle: A Case Study in Kolkata, India [26]

This is a machine-generated summary of:

Bera, Reema; Maitra, Bhargab: Commuters' Willingness-to-Pay for the Attributes of Plug-in Hybrid Electric Vehicle: A Case Study in Kolkata, India [26].

Published in: Transportation in Developing Economies (2021).

Link to original: https://doi.org/10.1007/s40890-021-00142-3.

Copyright of the summarized publication:

If you want to cite the papers, please refer to the original.

For technical reasons we could not place the page where the original quote is coming from.

Abstract-Summary

"The collected data were analyzed by developing multinomial logit (MNL) and mixed logit (ML) models to estimate overall commuters' WTP for a set of key attributes of PHEV namely purchase price, fuel cost reduction, battery range, public charging station, battery recharging time, battery warranty and tailpipe emission."

"In ML model, heterogeneity around the mean of random parameters with respect to sociodemographic characteristics such as age, monthly family income, car ownership, home-based parking facility and education, and trip characteristics such as average commuting trip length were examined, and WTP estimates were evaluated for different population sub-groups."

"The WTP estimates indicate that commuters perceive battery warranty as the major determinant influencing the choice of PHEV followed by battery recharging time and tailpipe emission."

"The potential users' income, availability of home-based parking facility, education, and average commuting trip length also substantially influenced their WTP for several attributes of PHEV in Indian context."

Introduction

"The widespread adoption of electric vehicles could contribute to the sustainability of transportation by decreasing reliance on conventional fuels (petrol or diesel) for transport, with the associated benefits of reduction in air pollution and GHG emissions [1]."

"The Government of India (GoI) has also developed strategies to push the adoption of electric vehicles."

"It is essential to gain an insight into the attributes influencing commuters' choice of electric vehicle, and evaluate them in terms of user perceived benefit such that appropriate manufacturing requirements and suitable government policies to increase the market share of electric vehicles could be identified."

"For a typical Indian context, the present investigation is carried out with reference to plug-in hybrid electric vehicle (PHEV)."

"By powering the vehicle through grid electricity, PHEV emits lower level of GHG emissions and air pollutants than CVs [2]."

Literature Review and Scope of the Research

"Qian and others [3] conducted stated choice experiment and developed discrete choice MNL and ML models to identify the impact of service and government policy attributes on consumer preferences for electric vehicles in China."

"Consumer preferences for electric vehicles depend on vehicle attributes (e.g. purchase price, driving range, fuel economy, charging time, maximum speed, acceleration, body type/size, CO_2 emission), service attributes (e.g. public charging station availability, service speed of public charging station, home charging facility), government policy attributes (e.g. subsidies, purchase tax relieves, free parking and bus lane access for electric vehicles, abolishing fuel subsidy and restricting age limit of conventional vehicles), individual characteristics (e.g. age, gender, income, car ownership, education, attitude towards alternative energy sources, environmental awareness) and trip characteristics (e.g. mileage, frequency of urban trips)."

Methodology

"The attributes and level identification, followed by generation of choice sets, survey design and data collection, and finally the selection of econometric models for choice analysis."

"Based on the review of past studies and discussion with subject experts, seven key attributes of PHEV namely purchase price [4, 5], fuel cost reduction [6, 7], battery range [8], public charging station [8, 9], battery recharging time [1, 9], battery warranty [8] and tailpipe emission [1, 9] were included in the SP choice experiment to analyze commuter preferences and WTP for these attributes."

"The attributes and their levels selected for the study were used to design hypothetical alternatives and choice sets."

"While the population statistics include all the urban residents of Kolkata, the sample data only consists of currents owners of CV."

"While database development for discrete choice modeling, all the quantitative attributes considered in the study namely purchase price, fuel cost reduction, battery range, public charging station, battery recharging time, battery warranty, and tailpipe emission were entered in cardinal linear form."

Model Specifications

"The mixed logit (ML) model is a powerful model that overcomes the limitations of MNL model by allowing more general substitution patterns and random taste heterogeneity across individuals' choices [10]."

"The model estimation in ML model is substantially influenced by the distributional assumption of the random parameters and number of draws considered in the simulation [11, 12]."

"For model estimation, the random parameters of ML models were assumed to follow constrained triangular distribution [13]."

"For number of draws to be considered in the simulation, Bhat [14] demonstrated that as compared to pseudo-random sequences that use random draws, non-random and more uniformly distributed sequences, known as "Halton" sequences, reduce the number of draws required for ML model estimation, reducing the computational time and/or simulation error associated with a given number of draws."

Results

"In Model ML2, Model ML3, Model ML4, and Model ML5, the WTP estimates for non-cost parameters corresponding to commuters with income $\leq$ 1,50,000 ₹/month, commuters with absence of home-based parking facility, commuters with education up to higher secondary and commuters with trip length $\leq$ 30 km respectively, were obtained by dividing the coefficient of random parameter with the coefficient of purchase price."

"Whereas, the WTP estimates corresponding to commuters with income > 1,50,000 ₹/month, commuters with presence of home-based parking facility, commuters who are graduate and above and commuters with trip length > 30 km in Model ML2, Model ML3, Model ML4, and Model ML5 respectively, were obtained by dividing the sum of coefficient of random parameter and heterogeneity around the random parameter with the coefficient of purchase price."

"No statistically significant heterogeneity was observed on mean estimates of battery warranty with respect to home-based parking facility, education, and average commuting trip length in Model ML3, Model ML4, and Model ML5, respectively."

Conclusions

"Substantially higher WTP estimates indicate that battery warranty and battery recharging time are the two most important attributes influencing commuters' choice to purchase PHEV in Indian context."

"It was interesting to note that the potential users' monthly family income, availability of home-based parking facility, education, and average commuting trip length also substantially influenced their WTP for several attributes of PHEV in Indian context."

"While the valuation of attributes in terms of WTP estimates provides useful insights into the attributes influencing commuters' choice to purchase PHEV, it also indicates the need for further investigation to analyze the mode shift from CV to PHEV by developing demand models."

"The authors would like to emphasize that even though this study has attempted to demonstrate a methodology to estimate commuter's WTP for the attributes of PHEV in a typical Indian context, the results are not tested for transferability."

8.2.2 *Accelerating Electric Vehicle Adoption: Techno-Economic Assessment to Modify Existing Fuel Stations with Fast Charging Infrastructure [27]*

This is a machine-generated summary of:

Ghosh, Nilanshu; Mothilal Bhagavathy, Sivapriya; Thakur, Jagruti: Accelerating electric vehicle adoption: techno-economic assessment to modify existing fuel stations with fast charging infrastructure [27].

Published in: Clean Technologies and Environmental Policy (2022).

Link to original: https://doi.org/10.1007/s10098-022-02406-x.

If you want to cite the papers, please refer to the original.

For technical reasons we could not place the page where the original quote is coming from.

Abstract-Summary

"With the increasing electric vehicle (EV) penetration, there arises an immediate need for charging infrastructure."

"We propose a novel methodology to assess the techno-economic feasibility of retrofitting an existing fuel station with EV charging infrastructure also known as Electric Vehicle Supply Equipment (EVSE)."

"To further enhance the value proposition, the potential of integrating Battery Energy Storage System (BESS) with EV charging infrastructure, which results in the reduction of grid connection costs, is studied."

"The configurations in this study are designed based on the technical considerations involved in retrofitting a typical fuel station as a fast charging facility for EVs."

"It is observed that the configurations with 4 EVSE, 1 BESS, and 8 h of operation and the configuration with 4 EVSE, 1 BESS, and 1 PV system for 8 h of operation are economically viable."

"The proposed methodology indicates that though the connection cost is the dominant factor affecting the feasibility, the use of BESS with or without PV can reduce the connection cost by almost 90% depending on the capacity of BESS."

"The methodology acts as a decision support tool to select a techno-economically feasible configuration of EVSE, BESS, and PV."

Introduction

"To reduce the dependency on the electricity grid, renewable energy systems like solar PV can be integrated into the EV charging stations (Colak and others [15]; Ilango and others [16])."

"Real-life scenario to assess the potential of implementing a fast charging station in an existing fuel station and the use of a renewable energy source (solar PV) to reduce the dependency on the electricity grid is missing in the existing literature. (Egbue and others [17]) have presented that lack of charging infrastructure and negative environmental impact owing to lack of renewables as key factors for not owning EVs among the respondents surveyed in the study."

"The potential of implementing a fast charging station in an existing fuel station and the use of a renewable energy source (solar PV) to reduce the dependency on the electricity grid has also been studied (Fernández and others [18]; Singh and others [19])."

Methodology

"The number of DCFC EVSE that can be implemented in a typical fuel station has been calculated considering the size of a typical fuel station (1200 sq."

"Calculation of the BESS capacity a. As the idea is to repurpose the existing fuel station without additional infrastructure cost, it is proposed to use the fuel storage tanks as the battery storage area."

"It can be safely considered that the ratio of energy supplied from the battery will be equal to the ratio of the power capacity of storage units (P_{bat} in kW) to the net power of the DCFCs (P_{DCFC} in kW)."

"A case study of a typical fuel station in the UK is considered and it will be able to accommodate 4 DCFCs."

Results and Discussions

"With an increase in the hours of operation, the NPV value becomes more positive as observed from the configuration 2C1S8h which has an NPV value of £93 k, IRR of 16% and a DPP of 6.3 years."

"With the increase in daily hours of operation the configurations 4C3S6h and 4C3S8h become economically viable with a highly positive NPV of £33 k and £165 k and IRR of 6% and 14%, respectively."

"When the hours of operation are increased, the configuration 3C1S1P6h has a positive NPV value of £44 k with an IRR of 6% and a DPP of 9.3 years, indicating that it is economically feasible."

"When the hours of operation are increased, the configurations 4C1S1P6h are only economically feasible with a positive NPV of £76 k IRR of 8% and DPP of 9.7 years."

Conclusion

"Techno-economic feasibility of retrofitting existing fuel stations with EVSE equipped with/without storage and with/without PV is analysed in this paper."

"The existing gas stations are ideally located to implement EV fast charging stations and the methodology proposed in this study presents a techno-economic feasibility method to repurpose the existing gas stations to install EV fast chargers."

"The installation of PV will reduce the dependency on the electricity grid and reduce the yearly fuel cost generated by using the electricity grid making the system more sustainable."

"The economic feasibility of PV-integrated fast charging stations will motivate investors to explore the potential of retrofitting existing fuel stations."

"There is potential for the work to be further expanded to include the revenue generated by batteries from participation in multiple electricity markets, which is likely to further improve the economic feasibility of battery integrated fast charging stations."

8.2.3 Techno-Economic Evaluation of Regulation Service from SEVs in Smart MG System [28]

This is a machine-generated summary of:

Siddiqui, Amira Nisar; Thomas, Mini S.: Techno-Economic Evaluation of Regulation Service from SEVs in Smart MG System [28]

Published in: Technology and Economics of Smart Grids and Sustainable Energy (2016).

Link to original: https://doi.org/10.1007/s40866-016-0016-z.

If you want to cite the papers, please refer to the original.

For technical reasons we could not place the page where the original quote is coming from.

Abstract-Summary

"The techno-economic analysis is based on simulation results obtained from a micro-grid test system operated with varying power generation and frequency regulation being provided by two different microgrid resources i.e., conventional battery storage system and proposed smart electric vehicles."

"The cost for providing frequency regulation using battery storage system and grid integration of vehicle is calculated and compared to find microgrid optimal operation option."

"The comparison of two costs reveals that smart electric vehicles have considerable economic potential for microgrid regulation service and hence provide better utilization of resources for optimal operation."

Introduction

"Microgrid is a localized grouping of electric power resources and loads that normally operate connected and synchronized with the traditional centralized grid (macro-grid), but can disconnect and function autonomously as physical and/or economic conditions dictate."

"Authors in [22] showed that microgrid system performance improved considerably when electric vehicles were used to help in frequency control and also that a large number of electric vehicles can be integrated while adopting advanced centralized electric vehicle charging control strategies without the need to proceed for grid reinforcement."

"Analysis is done to check the techno-economic viability of electric vehicle integration to enhance the microgrid operational efficiency."

"Smart electric vehicles are used to provide frequency regulation in place of battery storage system in a smart microgrid."

"The economic value of regulation service using vehicle to grid/ grid to vehicle (V2G/ G2V) operation of smart electric vehicles is compared with regulation provided by battery storage system."

Regulation Service Using SEVs

"An electric vehicle control strategy is designed and proposed here to enable electric vehicle to provide regulation-up (V2G) and regulation-down (G2V) service in response to the MG transient situations."

"The electricity from grid integration of vehicle (GIV) is costly when compared to electricity from large power plants but it can provide ancillary services like peak load shaving, regulation services, spinning reserves, etc. in a power distribution market."

"In case of excessive power generation or low power demand in the MG, frequency increase above the specified levels of vehicle controller and SEVs with SOC less than 100% move in G2V mode where controller initiates charging action and regulation-down service is aided."

"ΔP gives rise to frequency deviation Δf and for this kind of situation emergency regulation action of SEV is initiated through vehicle controllers to provide frequency regulation for the MG transient situation."

Simulation Results and Evaluation of Regulation Services

"The total generation from different sources of MG is varied from 50kW-500kW to analyze the effectiveness of SEVs operation for microgrid frequency regulation."

"Situation, 54.4% of SEVs move into V2G mode for providing MG frequency regulation with 40% SEVs which were initially in G2V mode either went into non-responsive mode or V2G mode depending upon their SOC."

"In 6b, with increase in MG generation from 271 to 350kW, 19.2% SEVs adopt G2V mode to provide frequency regulation."

"This cost is actually the revenue which MG utility has to pay to the SEV owners involved in regulation service."

"The results reveal that in a MG system, GIV for frequency regulation is cost effective."

"Results favours GIV and it is found that not only technically but economically also GIV ensures a good replacement of costly BSS in providing regulation services for the MG system."

Conclusion

"An analysis has been done to evaluate the use of electric vehicles for providing regulation in a microgrid system with intermittent power generation from distributed resources (PV array and wind turbine)."

"In smart microgrid environment an efficient regulation service can be provided using SEV."

"The results reveal that grid integration of electric vehicle has significant economic potential to provide regulation up and down without compromising on the driving schedule."

"Grid integration of vehicle greatly reduces the cost of ownership of electric vehicle and hence proves to be beneficial to the vehicle owner who receives revenue for providing regulation services without any extra expanses."

"The degradation cost of electric vehicle batteries is due to charging and discharging for two different services, hence this cost has to be separated into two components i.e. the cost due to battery degradation for mobility service of the vehicle and the cost due to power regulation service."

8.2.4 How Do Increases in Electric Vehicle Use Affect Urban Toll Ring Prices? [29]

This is a machine-generated summary of:

Krehic, Lana: How do increases in electric vehicle use affect urban toll ring prices? [29].

Published in: Public Choice (2022).

Link to original: https://doi.org/10.1007/s11127-022-01008-x.

If you want to cite the papers, please refer to the original.

For technical reasons we could not place the page where the original quote is coming from.

Abstract-Summary

"This paper assesses the effect of exempting electric vehicles from urban toll ring charges on the charge that is levied on conventional car drivers."

"Using panel data of Norwegian cities that have urban toll rings, I exploit regional variation in electric car adaption and find that owners of conventional cars pay 3.3 NOK (0.36 USD) more per passage due to the exemption."

"I find that local governments that are fragmented or have a left-wing majority increase toll charges more due to the loss in revenue."

"As the majority of electric vehicle owners have above-average incomes, the exemption implies a distributional effect in which low-income groups pay the largest portion of the increased toll price."

Introduction

"While toll charges are a means of raising funds for road construction and public transport, exempting electric vehicles from the charges reduces the potential revenues."

"The aim of this paper is to assess whether local governments raise urban toll ring charges to compensate for an increase in the number of exempted electric vehicles."

"An increase in toll charges to compensate for the exemption of high-income electric vehicle owners would be at the expense of middle-income and low-income households."

"I utilize data on the number of electric and conventional vehicles in cities with urban toll rings and their adjacent municipalities from 2010 to 2017, which is a period that coincided with increased electric car ownership."

"A back-of-the-envelope calculation based on estimates in this paper suggests that in 2017, a conventional vehicle owner paid 3.3 NOK (0.36 USD) more every time she crossed an urban toll ring compared with the counterfactual case in which electric vehicles were not exempted from the charge."

Related Literature

"This paper, in addition to studying how local governments in general respond to revenue shortfalls, explores how the political composition of local governments can produce heterogeneous outcomes in the toll pricing policy."

"Despite the weak relationship between party ideology and local economic policy found in the United States (see for example Reed [23] or Potrafke [24]), Pettersson-Lidbom [25] notes that higher shares of left-wing representatives in Swedish local councils lead to both greater spending and higher taxation."

"The existing literature therefore implies that toll prices are potentially higher in municipalities where the left-wing share is larger."

"An additional factor that supports this perception is that left-wing parties in Norway are more positive towards the use of toll charges to finance road construction."

"It is not given that left-wing majority municipalities are more likely to increase toll prices when faced with declining revenues."

Institutional Background

"Every month, the local toll company sends an invoice to the vehicle's registration address, where the number of passes, toll price, and rebates are summed."

"Urban toll rings still serve as a mechanism for raising revenue for road infrastructure investments, some of which operate similarly to a congestion charge with rush-hour charges."

"Two types of price increases are possible during the urban toll ring's operation period."

"Municipalities consider increasing the toll price when additional revenue is needed either to add projects to the city package or if the planned projects are underfinanced."

"In some cases, the toll companies reduced the maximum discount instead of increasing the toll price."

"I estimate the effect of the electric vehicle share on the urban toll ring price under the assumption that the price commuters face is the price that corresponds to the maximal discount."

Data

"I draw data from Statistics Norway on the number of registered cars in all 419 Norwegian municipalities by fuel type."

"Since the data are provided on an annual basis, the prices of the toll rings are weighted annual averages, where the weights are the numbers of days for which the corresponding prices are in operation."

"During the sample period there are eight cities with toll rings, which suggests that only eight municipalities should be included in the analysis."

"I include municipalities that have strong proximity and economic and labor market attachments to the municipality where the toll ring is located."

"If a city has an urban toll ring, the inhabitants of neighboring municipalities will be assumed to face the same price as the inhabitants of the ring city."

Empirical Approach

"I estimate the following equation: where $\ln(\text{realprice})_{it}$ is the natural logarithm of the inflation-adjusted price of a toll ring in municipality i in year t, and ShareEV_{it} is the number of electric vehicles divided by all vehicle types in municipality i at time t. An alternative approach would be to have the total number instead of electric vehicle share as the main variable of interest."

"The key parameter of interest is α_1, which provides an estimate of the effect of an increase in the share of electric vehicles on the price of the toll ring."

"The share of electric vehicles variable is implicitly lagged so that politicians observe the number of electric vehicles at the end of last year before deciding to change the road toll price."

"The approach should be in line with the exclusion criteria because the share of electric vehicles in the distant municipalities does not influence the toll ring prices in the focal city."

Results

"I start this section by estimating the relationship between electric vehicle share and urban toll ring prices with a fixed-effects approach."

"The estimate from the fixed-effects model suggests that an increase in the share of electric vehicles of one percentage point, for example from 10 to 11%, will increase the toll ring price by approximately 4.2%."

"A 10-percentage-point increase from a 10% and 50% electric vehicle share, respectively, might have different effects on the toll ring price."

"The point estimate of the squared variable is negative, which implies a concave relationship between electric vehicle share and toll price."

"One would expect that if the total number of cars increases, while controlling for the share of electric vehicles, the potential revenue of the toll ring is larger and the toll price should be lower."

"Because the estimates do drop somewhat, the toll price has a positive effect on the electric vehicle share."

Political Heterogeneity

"A higher effective number of parties (ENoP) represents less political strength and is expected to increase the toll price."

"The baseline effect suggests that in cases where municipalities are governed by coalitions, the effect of electric vehicles on toll price is much stronger compared with the main estimates."

"In councils with a right-wing majority, the effect of an increase in electric vehicle share on the real toll price is close to zero."

"This might imply that right-wing governments, because of reduced toll revenues, reduce road expenditure instead of increasing the toll charge."

"The results in column (2) suggest that governments with left-wing majorities are more prone to increasing the toll price than right-wing majorities, which is in line with the stated ideology of left-wing parties."

"However, that the increase in toll prices by left-wing majorities is lower than the price increases performed by governments with no majority."

Distributional Effects

"I explore how the ownership of electric and conventional vehicles varies by income group to assess whether some income groups receive a net benefit from the policy."

"How much more is a conventional vehicle owner paying per urban toll ring passage as a consequence of the larger electric vehicle share?"

"This naïve calculation suggests that a conventional vehicle owner pays 3.3 NOK (0.36 USD) more per toll ring crossing today than in the counterfactual scenario in which there is no exemption policy."

"In section B, I calculate the value of the excess payment by the conventional car owner to bridge the toll company's revenue gap according to income group."

"The increased toll price is a burden that is shifted mostly onto low-income non-electric vehicle owners, thereby resulting in an inverse relationship between cost and income."

Concluding Remarks

"Many schemes exempt less-polluting vehicles from charge to obtain a cleaner car fleet, which might offset some of these effects and may have implications for social welfare."

"This paper explores how toll charges, set by local governments, are affected when selected vehicle types are exempt from the charges."

"By analyzing the prices in several urban toll rings in Norway over time, I find that an increase in the share of electric vehicles increases the toll price."

"I find that the increase in toll charges is shifted onto the below-average income group, implying a negative distributional effect."

"The total welfare effect of the exemption of electric vehicles from toll charges is uncertain."

"If the revenue from urban toll charging is recycled so that it benefits the lower-income groups more, such as investment in public transportation, this will have a positive effect on the total welfare."

8.2.5 Energy Trading for Electric Vehicle Aggregator Under Random Participation of EV Owners [30]

This is a machine-generated summary of:

Kannan, Anuradha; Tumuluru, Vamsi Krishna: Energy trading for electric vehicle aggregator under random participation of EV owners [30]

Published in: Energy Systems (2022).

Link to original: https://doi.org/10.1007/s12667-022-00536-1.

If you want to cite the papers, please refer to the original.

For technical reasons we could not place the page where the original quote is coming from.

Abstract-Summary

"Different from the existing works, the proposed model considers the randomness in the charging and discharging preferences of the EV owners on the operating day."

"Such uncertainty arises because the EV owners may randomly decide to participate on the operating day."

"Further, the EVA faces other uncertainties related to both the market-level data (i.e., electricity prices, real-time dispatch signal) and the fleet-level data (e.g., energy consumption of EVs, arrival and departure times) prior to participating in the electricity markets."

"The performance of the proposed model is analyzed by comparing with the traditional model which ignores the randomness in the participation of the EV owners."

Introduction

"In a realistic setting, every EV owner may not actively participate in energy trading and instructed real-time regulation service on each operating day (i.e., the day after submitting energy bids to the DAM)."

"Unlike the existing works, in this paper, the EVA performs two-way energy trading (buy and sell) and up-regulation commitments in the DAM without knowing the actual number of active EV owners on the operating day."

"The proposed formulation ensures that the EVA offers up-regulation capacity only when the real-time dispatch signal can be strictly followed by the active EV owners under any random scenario."

"Proposes a two-stage stochastic MIP that allows an EVA to buy, sell and commit up-regulation capacity in the DAM despite facing the uncertainty about the active status of the EV owners on the operating day (i.e., the day after the DAM)."

System Model

"The EVA performs two-way energy trading (i.e., buy and sell power) on behalf of the active EV owners by submitting price-taking bids in the DAM and the RTMs."

"The EVA also commits capacity for up regulation in the DAM at each location only when it can strictly follow the real-time dispatch signal under all random scenarios."

"The EVA receives payments from the market operators for the energy it sells in the DAM and RTMs."

"Further, the EVA also receives payments from the DAM operator for committing up-regulation capacity and providing instructed real-time instructed regulation service on the operating day."

"Unlike the existing works, the main challenge before the EVA is the uncertainty about the active status of the EV owners on the operating day."

Proposed Energy Trading Problem

"This section presents the formulation and solution method for the proposed energy trading problem."

"The second summation represents the revenue made by the EVA for selling power in the DAM."

"The second summation represents the revenue made by the EVA for selling power in the RTM."

"We simplify the representation of the proposed formulation in order to explain the solution method."

Results and Discussion

"This was because as the uncertainty in participation of the EV owners increased, the EVA became more conservative in the DAM."

"This resulted in lesser revenue in the DAM and thus the total expected net cost increased."

"It can be seen that each fleet achieved a lower expected net cost under the proposed model compared to the traditional model."

"It was observed that the expected net cost depended on the number of active EV owners under the fleets."

"The main contribution to the lower expected net cost under the proposed model was the higher revenue obtained in the DAM and RTMs compared to the traditional model."

"Further, by taking the uncertainty into account, the EVA was able achieve better revenue in the RTMs under the proposed model."

"It may be noted that while the revenue of the EVA increased, the expected net cost per fleet did not increase significantly."

Conclusion

"This paper proposed a novel two-stage stochastic two-way energy trading problem for an EVA for tackling the problem of random participation by the EV owners on the operating day."

"Further, owing to such random participation, the charging and discharging preferences of the EV owners also became random variables."

"The EVA committed up regulation capacity in the DAM based on the random charging and discharging preferences of the EV owners."

"The results showed that the proposed model outperformed the traditional model despite variation in the number of active EV owners and their random charging and discharging preferences."

Bibliographys

1. Nie Y, Wang E, Guo Q, Shen J (2018) Examining shanghai consumer preferences for electric vehicles and their attributes. Sustainability 10(6):2036. https://doi.org/10.3390/su10062036
2. Krupa JS, Rizzo DM, Eppstein MJ, Brad Lanute D, Gaalema DE, Lakkaraju K, Warrender CE (2014) Analysis of a consumer survey on plug-in hybrid electric vehicles. Transp Res Part A Policy Pract 64:14–31. https://doi.org/10.1016/j.tra.2014.02.019
3. Qian L, Grisolía JM, Soopramanien D (2019) The impact of service and government-policy attributes on consumer preferences for electric vehicles in China. Transp Res Part A Policy Pract 122:70–84. https://doi.org/10.1016/j.tra.2019.02.008
4. Potoglou D, Kanaroglou PS (2007) Household demand and willingness to pay for clean vehicles. Transp Res Part D Transp Environ 12:264–274. https://doi.org/10.1016/j.trd.2007.03.001
5. Hsu CI, Li HC, Lu SM (2013) A dynamic marketing model for hybrid electric vehicles: a case study of Taiwan. Transp Res Part D Transp Environ 20:21–29. https://doi.org/10.1016/j.trd.2013.01.001

6. Tanaka M, Ida T, Murakami K, Friedman L (2014) Consumers' willingness to pay for alternative fuel vehicles: a comparative discrete choice analysis between the US and Japan. Transp Res Part A Policy Pract 70:194–209. https://doi.org/10.1016/j.tra.2014.10.019
7. Hackbarth A, Madlener R (2013) Consumer preferences for alternative fuel vehicles: a discrete choice analysis. Transp Res Part D Transp Environ 25:5–17. https://doi.org/10.1016/j.trd.2013.07.002
8. Higgins CD, Mohamed M, Ferguson MR (2017) Size matters: how vehicle body type affects consumer preferences for electric vehicles. Transp Res Part A Policy Pract 100:182–201. https://doi.org/10.1016/j.tra.2017.04.014
9. Byun H, Shin J, Lee CY (2018) Using a discrete choice experiment to predict the penetration possibility of environmentally friendly vehicles. Energy 144:312–321. https://doi.org/10.1016/j.energy.2017.12.035
10. McFadden D, Train K (2000) Mixed MNL models for discrete response. J Appl Econom 15(5):447–470. https://doi.org/10.1002/1099-1255(200009/10)15:5%3C447::AID-JAE570%3E3.0.CO;2-1
11. Baker RP, Bradburn NM, Johnson RA (1995) Computer-assisted personal interviewing: an experimental evaluation of data quality and cost. J Off Stat 11(4):413–431
12. Train KE (2009) Discrete choice methods with simulation. Cambridge University Press
13. Hensher DA, Rose JM, Greene WH (2015) Applied choice analysis, 2nd edn. Cambridge University Press
14. Bhat CR, Eluru N, Copperman RB (2007) Flexible model structures for discrete choice analysis. In: handbook of transport modelling. Emerald Group Publishing Limited, pp 75–104
15. Colak I, Bayindir R, Aksoz A, Hossain E, Sayilgan S (2016) Designing a competitive electric vehicle charging station with solar PV and storage. In: INTELEC, International telecommunications energy conference (Proceedings), 2016-September. https://doi.org/10.1109/INTLEC.2015.7572480
16. Ilango R, Vengadachalam N, Seethalakshmi VS (2022) Charging demand based on the interaction among electric vehicles and renewable energy sources using hybrid technique. Clean Technol Environ Policy 2022:1–20. https://doi.org/10.1007/S10098-022-02334-W
17. Egbue O, Long S, Samaranayake VA (2017) Mass deployment of sustainable transportation: evaluation of factors that influence electric vehicle adoption. Clean Technol Environ Policy 19(7):1927–1939. https://doi.org/10.1007/S10098-017-1375-4/TABLES/5
18. Fernández G, Torres J, Cervero D, García E, Alonso MÁ, Almajano J, Machín S, Bludszuweit H, Sanz JF (2019) EV charging infrastructure in a petrol station, lessons learned. In: 2018 international symposium on industrial electronics, INDEL 2018—Proceedings, pp 1–6. https://doi.org/10.1109/INDEL.2018.8637635
19. Singh A, Shaha SS, Nikhil PG, Sekhar YR, Saboor S, Ghosh A (2021) Design and analysis of a solar-powered electric vehicle charging station for indian cities. World Electr Veh J 12(3):132. https://doi.org/10.3390/WEVJ12030132
20. Di W, Aliprantis DC, Ying L (2012) Scheduling and dispatch for aggregators of Plug-In EV. IEEE Trans Smart Grid 3(1):368–376
21. Xenias D, Axon CJ, Whitmarsh L, Connor PM, Balta-Ozkan N, Spence A (2015) UK Smart Grid development: an expert assessment of the benefits, pitfalls and functions. Renew Energy 81:89–102. https://doi.org/10.1016/j.renene.2015.03.016
22. Lopes JA, Soares FJ, Rocha Almeida PM (2011) Pecas Integration of electric vehicles in the electric power system. Proc IEEE 1:99
23. Reed WR (2006) Democrats, republicans, and taxes: evidence that political parties matter. J Public Econ 90(4–5):725–750. https://doi.org/10.1016/J.JPUBECO.2004.12.008
24. Potrafke N (2017) Government ideology and economic policy-making in the United States—a survey. Public Choice 174(1):145–207. https://doi.org/10.1007/S11127-017-0491-3
25. Pettersson-Lidbom P (2008) Do parties matter for economic outcomes? A regression discontinuity approach. J Eur Econ Assoc 6(5):1037–1056. https://doi.org/10.1162/JEEA.2008.6.5.1037

26. Bera R, Maitra B (2021) Commuters' willingness-to-pay for the attributes of plug-in hybrid electric vehicle: a case study in Kolkata, India. Transp Dev Econ. https://doi.org/10.1007/s40890-021-00142-3
27. Ghosh N, Mothilal Bhagavathy S, Thakur J (2022) Accelerating electric vehicle adoption: techno-economic assessment to modify existing fuel stations with fast charging infrastructure. Clean Technol Environ Policy. https://doi.org/10.1007/s10098-022-02406-x
28. Siddiqui AN, Thomas MS (2016) Techno-economic evaluation of regulation service from SEVs in smart MG system. In: Technology and economics of smart grids and sustainable energy. https://doi.org/10.1007/s40866-016-0016-z
29. Krehic L (2022) How do increases in electric vehicle use affect urban toll ring prices? Public Choice. https://doi.org/10.1007/s11127-022-01008-x
30. Kannan A, Tumuluru VK (2022) Energy trading for electric vehicle aggregator under random participation of EV owners. Energy Syst. https://doi.org/10.1007/s12667-022-00536-1

Chapter 9
Skills and Workforce Training

9.1 Introduction by Editor

9.1.1 Introduction

The transition from ICE to EVs will require changes in the workforce. Certain skills are required in ICE which may become redundant in EVs. This chapter will discuss the skills required in EV assembly and manufacturing, and repair. It will examine the training necessary for inculcating this skill set in workers. The chapter will also discuss what the EV transition will mean for the country in terms of the workforce engaged in ICE versus EVs.

9.1.2 Jobs in the EV Sector

The EV transition will require the skilling and reskilling of workers due to the differences between the ICEVs and EVs. The skills required for the EVs need to be seen in the context of the parts that go into each vehicle. The jobs can be classified into [1]:

(1) **battery manufacturing/assembling**—these include jobs in battery manufacturing, battery management system, manufacturing and assembly of components, vehicle body manufacturing, and assembly line [2].
(2) **charger manufacturing**—jobs include manufacturing chargers.
(3) **wholesale installation of charging points**—jobs in the installation of all kinds of chargers.
(4) **charge point maintenance**—jobs in charging operations management. Though there will be a lower overall requirement for maintenance in EVs, these jobs are critical also since these are more labour-intensive than other operations [3].

S. Ray and D. Mukherji (eds.), *A Primer on Electric Vehicles in India*,
https://doi.org/10.1007/978-981-97-8966-5_9

(5) **electricity generation**—grid connection jobs as well as those in RE integration [4].

Taking the entire value chain of the EV into consideration, the jobs associated with parts that are found in both ICE and EV will continue. These include start-stop systems, pistons, spark plugs, turbochargers, clutch/torque control. However, other jobs will be impacted as the transition progresses [5]. Engine parts, gearboxes, and exhaust pipes are not required in EVs. 50% of the automobile industry is engaged in engine manufacturing and these jobs are at risk [6, 7]. The new parts in EVs include the DC/DC converter, inverter, power distribution module, onboard charger, battery management system (BMS), vehicle interface control module (VICM), EV communication controller (EVCC), and a simplified gearbox.

9.1.3 Skills Required in the EV Sector

The skills required [8] in the EV sector correspond to the jobs that are new due to the nature of the vehicles and hence are different from the ICE sector. It is useful to categorize the jobs and the skills in the following way:

(1) **For R&D** [9, 10]—in R&D, there are jobs associated with vehicle transmission, design, vehicle testing, software, and battery technology. The skills required to do these functions include electrical, electronics, computer, and mechanical engineering, etc. There are overlaps with ICE in the skill sets for aerodynamics (in the design segment), driver monitoring (in the vehicle testing segment), autonomous driver-assisted vehicle technology, or AI-enabled technology (in the software segment). However, skills needed to examine battery safety in the vehicle testing segment (i.e. electrical engineering), or those related to battery technology (e.g. in efficient power usage, battery composition, and end-of-life management) are not required in ICE vehicles. The skills required in these cases are electrical, electronics, chemical engineering, and supply chain management, respectively.

(2) **For manufacturing**—the processes involved are the manufacture and assembly of components, vehicle body manufacture/assembly, manufacturing of batteries, and battery management system (BMS). The skill set involved in each case is electrical/electronic engineering, vehicle system engineering, chemical, and mechanical engineering. This is true of many parts which are not found in ICE vehicles. These functions can also be performed by Not in Employment (NEET) and Technical and Vocational Education and Training (TVET) pass-outs [11].

(3) **For Sales**—The features of an EV need to be understood to make sales. For this trained TVET workforce is needed [12].

(4) **For Services**—service technicians are needed for battery and motor related to EVs, operational digital interface, and vehicle safety. The requirement in this segment is people with skills to handle the above, and NEET, and TVET

trained. The training and reskilling of service technicians for diagnostics on EV performance is also important [13].

(5) **For End of Life**—the jobs include collection, battery testing, segregation, recycling, and safety in disposal. Some of these functions can be performed by electrical engineers or NEET, TVET-trained personnel, while battery testing can be performed only by engineers and TVET-trained people.

An estimate of the number of jobs associated with each of the segments above is helpful in understanding the skillsets that are in or likely to be in short supply and upskilling [14] required.

9.1.4 Gaps in Training

There are many courses that have been designed to address the engineering requirements of the sector [15]. Many OEMs have in-house training where skills are imparted [16]. The gaps in training are particularly a problem for tier 3 and 4 auto component manufacturers [17]. There is an apprenticeship programme in the country including the National Apprenticeship Promotion Scheme (NAPS), National Employment Enhancement Mission (NEEM), and Employability Enhancement Training Program (EETP) [18]. There is a need to ASDC to develop programs for upskilling especially in the NEET segment.

9.1.5 Way Head

There has been a discussion about whether the EV transition will lead to job losses. The auto industry employed 32 million people in 2018 and is expected to employ 45 million in 2026. How many of those jobs will be in the EV sector? To answer this question, an estimate of the jobs in each segment of the value chain is needed. At another level, according to a report by OMI (2022), 1.2 lakh new jobs will be created (assuming a 30% share of EVs in 2030 [19]). The issue of the skill gap can also be addressed in a similar way, but information can be obtained by mapping the skill requirement in each segment with the number of people available to fulfill it. NASSCOM (2019) estimates that 15 lakh engineering graduates pass out each year. Data from Statista (2021) shows that in terms of students enrolled in engineering at an undergraduate level, the lowest is in electrical engineering at 354.49 (in 1000s).

To summarise, as the industry evolves industry-specific skills would be required in the following domains:

Energy—battery, BMS, and charging infrastructure requiring chemistry and electric engineering knowledge.

Electrical and Mechanical—motor, gear reducer, and drive lines requiring electric and mechanical engineering knowledge.

Electronics and IT—software, connectivity, and data skills.

So, are we likely to face a shortage of electrical engineers as the EV sector evolves? Without more granular data, this question cannot be answered. Considering the entire skill sets required in the sector with the expected growth of the EV sector can help answer the question.

Machine Generated

Kannayeram et al. (2023) proposes a new hybrid system to analyze the impact of the electric-vehicle-charging-station with the combination of renewable energy and grid-connected system. Electric vehicle charging stations can also provide local economic benefits by creating jobs and supporting the growth of the electric vehicle industry.

References

Bhattacharjee, A. (2022) Clean Mobility Shift. 27th December. Link: https://cleanmobilityshift.com/industry/is-indias-automotive-workforce-ready-for-an-ev-future/

E & Y (2019) Human resources and skill requirements in the automotive sector (2026).

GIZ (2021) Status quo analysis of various segments of electric mobility and low carbon passenger road transport in India.

Mckinsey (2022) The future of mobility: Transforming to be ahead of the opportunity. Report Link: https://www.mckinsey.com/industries/automotive-and-assembly/our-insights/the-future-of-mobility-transforming-to-be-ahead-of-the-opportunity

Munjal, N. (2022) India's electric vehicle transition: Impact on the auto industry, Economic Times, April 5, 2022. Link: https://economictimes.indiatimes.com/industry/auto/two-wheelers-three-wheelers/indias-electric-vehicle-transition-impact-on-the-auto-industry/articleshow/90666775.cms?from=mdr

NASSCOM (2019) Cloud Next Wave of Growth In India. Report. New Delhi.

OMI (2022) *Skilling Indians for an EV-Ready World, February 2022; Ola Mobility Institute.*

Ray, S., and Thakur, V., (2022) The Challenge of Skill gaps in the Indian EV sector. Chapter 2. In Moerenhout, T., Goldar, A., Ray, S., Shingal, A., Goel, S., Agarwal, P., Jain, S., Thakur, V., Tarun., Gaurav, K., Aneja, DA., Bansal, A. "Understanding Investment, Trade, and Battery Waste Management Linkages for a Globally Competitive EV Manufacturing Sector." Report. New Delhi. 2022. https://icrier.org/publications/understanding-investment-trade-and-battery-waste-management-linkages-for-a-globally-competitive-ev-manufacturing-sector/

Rizvi, K. and P. B. Tiwari (2019) Industrial Revolution 4.0 and the Impact on Automotive Sector. Friedrich Ebert Stiftung.

Soman, A., Ganesan, K., and H. Kaur (2019) India's Electric Vehicle Transition: Impact on the Auto Industry and Building the EV Ecosystem, CEEW and Shakti Sustainable Energy Foundation report.

Statista (2021) Number of students enrolled in engineering at an undergraduate level across India in 2021, by discipline. Link: https://www.statista.com/statistics/765482/india-number-of-students-enrolled-in-engineering-stream-by-discipline/

TERI (2019) Faster adoption of electric vehicles in India: Perspective of consumers and industry. New Delhi.

WRI (2022) Just Transition And Skill Development In The Electric Vehicle Industry, Conference Proceedings, Link: https://www.wri.org/research/just-transition-and-skill-development-electric-vehicle-industry

Yarlagadda, K. (2023) India's Transition To EVs Threatens Millions Of Auto Sector Jobs, At A Time Of Rising Nationwide Unemployment, Article 14. 29 May 2023. https://article-14.com/post/india-s-transition-to-evs-threatens-millions-of-auto-sector-jobs-at-a-time-of-rising-nationwide-unemployment--64740b579a7ce

Footnotes/Endnotes

[1] Rizvi and Tiwari (2019).

[2] OMI (2022)

[3] https://cleanmobilityshift.com/industry/is-indias-automotive-workforce-ready-for-an-ev-future/

[4] OMI (2022)

[5] https://economictimes.indiatimes.com/industry/auto/two-wheelers-three-wheelers/indias-electric-vehicle-transition-impact-on-the-auto-industry/articleshow/90666775.cms?from=mdr

[6] TERI (2019)

[7] Other jobs that are likely to be impacted include mechanics. See: https://article-14.com/post/india-s-transition-to-evs-threatens-millions-of-auto-sector-jobs-at-a-time-of-rising-nationwide-unemployment--64740b579a7ce

[8] WRI (2022)

[9] OMI (2022)

[10] Ray and Thakur (2022).

[11] OMI (2022)

[12] Ibid.

[13] https://cleanmobilityshift.com/industry/is-indias-automotive-workforce-ready-for-an-ev-future/

[14] https://www.mckinsey.com/industries/automotive-and-assembly/our-insights/the-future-of-mobility-transforming-to-be-ahead-of-the-opportunity

[15] Ibid.

[16] https://cleanmobilityshift.com/industry/is-indias-automotive-workforce-ready-for-an-ev-future/

[17] ibid.

[18] E&Y (2019)

[19] Soman et al. (2019).

9.2 Machine Generated Summaries

Disclaimer: The summaries in this chapter were generated from Springer Nature publications using extractive AI auto-summarization: An extraction-based summarizer aims to identify the most important sentences of a text using an algorithm and uses those original sentences to create the auto-summary (unlike generative AI). As the constituted sentences are machine selected, they may not fully reflect the body of the work, so we strongly advise that the original content is read and cited. The auto generated summaries were curated by the editor to meet Springer Nature publication standards. To cite this content, please refer to the original papers.

Machine generated keywords: method, complexity, propose method, renewable energy, peak, charge station, artificial, coordination, system propose, type charge, station, propose, power grid, vehicletogrid, voltage.

9.2.1 Impacts of Electric Vehicle Charging Station with the Integration of Renewable Energy with Grid Connected System: A Hybrid Technique [7]

This is a machine-generated summary of:

Kannayeram, Ganapathiapillai; Muniraj, Rathinam; Saravanan, Ragavan: Impacts of electric vehicle charging station with the integration of renewable energy with grid connected system: a hybrid technique [7].

Published in: Clean Technologies and Environmental Policy (2023).

Link to original: https://doi.org/10.1007/s10098-023-02548-6.

If you want to cite the papers, please refer to the original.

For technical reasons we could not place the page where the original quote is coming from.

Abstract-Summary

"This manuscript proposes a new hybrid system to analyze the impact of the electric-vehicle-charging-station with the combination of renewable energy and grid-connected system."

"The proposed method is the implementation of Forensic-Based Investigation (FBI) and Artificial Transgender Longicorn Algorithm (ATLA) called FBI–ATLA method."

"While type 1 charging stations utilize the power grid, renewable energy sources and vehicle-to-grid technology, type 2 charging stations have swappable batteries."

"The charging station may be met by four dissimilar power sources: (i) power grid (PG); (ii) batteries stored in charging stations; (iii) renewable energy sources (e.g., solar-energy); (iv) the energy discharged from electric vehicles (EV) to grid via vehicle-to-grid (V2G) connection."

"The implementation of the FBI–ATLA introduced through the MATLAB platform, and the performance of proposed method is compared with existing methods."

"The result shows that the proposed system is proficient than other existing techniques."

"This manuscript proposes a new hybrid system to analyze the impact of the electric-vehicle-charging-station (EVCS) with the combination of renewable energy and grid-connected system."

Introduction

"The advanced approaches, such as phases of joint coordination and DG dispatching network, use an improved state of charge of a group of EVs from a sensitive position as the secondary control level and a constant electric-vehicle state of charge as the primary control level."

"In Mehta and others [1], a synchronous charging of electric vehicle is proposed to reduce the distribution system load variance."

"The role of distribution transformers on smart networks using synchronous charging of electric vehicles was explored at Jimenez and Garci [2]."

"In Wenge and others [3], the synchronous charging of hybrid electric vehicle decrease the loss of distribution network was studied."

"The research hypotheses are the combination of renewable energy with grid-connected electric vehicle charging stations to provide the computational complexity, reduce the cost, improve efficiency and enhance the network power."

Recent Research Work: A Brief Review

"Khalkhali and Hosseinian [4] have advised novelty model as well as operates the instruments of EV parking lots such as charging modes are utilized to change from level-1 to level-2."

"A technique called three-stage scheduling platform is based on SP and MPC introduces a real-time periods such as periodic and intra are planned for undertaking the joint scheduling of unbendable loads, quick and slow-charged vehicles."

"In these technique, energy was given to the EVs of different charging classes in order to prevent or minimize the need for discrete quick charging stations in metropolitan regions."

"Amini and others [5] have suggested a hierarchical method PC technique for eco-cooling the autonomous vehicle to minimize the EC via real-time prediction as well as the multi-time scale optimization."

"Recent reviews portray multi-objective optimization issues that are coordinate with EVs as well as DGs together in an imbalanced DG electric vehicle is an important contribution factor."

System Model Configuration

"A 2-way association is made in the middle of PG, and EV charging station created to get electric power from the grid as well as back to the grid via B–G. Two types of charging station are utilized based on supplying the electricity capacitors."

"In the 1st phase, system minimizes the locating costs of charging station prior to the electric vehicle power demand."

"The 2nd phase decision involves decisions for dispatching the power in between the charged stations as well as to EV."

"The remaining 2 term of equation denotes the cost values, which is not yet satisfy electric demands and storage battery power station."

Proposed Hybrid Technique

"This section explains the proposed hybrid system, which is the combination of both FBI and Artificial Transgender Longicorn Algorithm (ATLA), named as Forensic-Based ALTA Technique."

"After using a step-by-step arbitrary search, the selected male longicorn transmits sex using maximum fitness based on fitness function (Han and others [6])."

"When the male longicorn is exposed to sex pheromone during sexual attraction, they move toward the female longicorn by indicating the source of the enrichment aswhere the i-th male position is denoted as X_i; the current number of iteration is represented as 't'; the position of female longicorn with maximal fitness in the mob indicates X_{best}."

"ATLA is used to calculate the fitness value for each longicorn."

"The sex of the female long corn receives the pheromone at the male longicorn threshold value, while the outer threshold value of the male longicorn engages in spontaneous search."

"The male longicorn in the rating value is affected through the sex pheromone."

Results and Discussion

"The impact of electric vehicles charge station (EVCS) utilize an integrates of solar energy with network connected system is analyzed by using a hybrid technique called FBI–ATLA."

"The method is greatly receptive to a count of time-dependent factors, like solar power obtain ability, vehicle and electricity prices flows on cell i of specified year t. For example, it is viewed that energy demand of EVs is mainly met through the network and V2G through low-cost hours of operation as well as unavailability of solar power is from 8:00 p.m. to 8:00 a.m. Then again, the demand is met initially through V2G and solar-power grid through peak hours of operation (i.e., from 10:00 a.m. to 2:00 p.m.) and solar power availability."

"EVC opens an additional charging station while the solar power accessibility decreases as of 50%."

Conclusion

"The effect of EVCS on power grid is analyzed by utilizing a hybrid technique named as FBI–ATLA technique."

"The FBI–ATLA technique is used to solve upgrade problems in the system and to solve the common power grid costs of the electric current associated with the creation of a charging station by allowing the decision maker efficiently supply the electricity demand."

"Electric vehicle charging stations can also provide local economic benefits by creating jobs and supporting the growth of the electric vehicle industry."

"The availability and reliability of renewable energy sources can be affected by weather conditions, which could impact the charging stations performance and consistency."

"The cost of building and maintaining charging stations and integrating them with renewable energy sources and V2G technology might be higher than traditional gas stations."

Bibliography

1. Mehta R, Srinivasan D, Khambadkone AM, Yang J, Trivedi A (2018) Smart charging strategies for optimal integration of plug-in electric vehicles within existing distribution system infrastructure. IEEE Trans Smart Grid 9(1):299–312
2. Jimenez A, Garci N (2011) Power flow modeling and analysis of voltage source converter-based plug-in electric vehicles. In: Power and energy society IEEE general meeting, 2011 IEEE
3. Wenge C,Arendarski B, Haensch K, Naumann A, Komarnicki P (2012) Electric vehicle simulation models for power system applications. In: 2012 IEEE power and energy society general meeting conference
4. Khalkhali H, Hosseinian SH (2020) Multi-stage stochastic framework for simultaneous energy management of slow and fast charge electric vehicles in a restructured smart parking lot. Int J Electr Power Energy Syst 116:105540
5. Amini MR, Kolmanovsky I, Sun J (2020) Hierarchical MPC for robust eco-cooling of connected and automated vehicles and its application to electric vehicle battery thermal management. IEEE Trans Control Syst Technol 29(1):316–328
6. Han X, Du X, Yu P (2020) ATLA: A novel metaheuristic optimization algorithm inspired by the mating search behavior of longicorn beetles in the nature. IOP Conf Ser Mater Sci Eng 782(5):052028
7. Kannayeram, Ganapathiapillai; Muniraj, Rathinam; Saravanan, Ragavan Impacts of electric vehicle charging station with the integration of renewable energy with grid connected system: a hybrid technique. Clean Technologies and Environmental Policy (2023). https://doi.org/10.1007/s10098-023-02548-6

Part V
Looking Ahead

The final part of the book examines the way ahead, the geopolitics behind mineral sourcing, and the path for EV transition in Chapter 10.

Chapter 10
Conclusion and Way Ahead

10.1 Introduction by Editor

10.1.1 Introduction

Everything you wanted to know about electric vehicles in the Indian context has been discussed in the book. It takes a value chain approach by discussing upstream and then downstream activities in the battery value chain. It discusses the ecosystem and the policies and regulations governing EVs, and financing the EV transition. It discusses skills and jobs in the EV value chain. This chapter brings together all the strands explored in the book by way of a conclusion. It will discuss some possible pathways for the EV transition in India. The pathways will depend on a host of factors including geopolitics.

10.1.2 Discussion in the Book on Key Themes

The key takeaways from the book are:

1. Understanding EVs—there is a big paradigm shift in the automotive industry
2. Supply chain challenges for battery raw materials remain and so do technology gaps. How the Indian government and industry can work together to overcome these challenges is the question
3. Mapping out the scale of opportunities for Indian companies, both for domestic and export markets in the EV era. If the above-mentioned challenges are met, it can also be an opportunity for the industry

S. Ray and D. Mukherji (eds.), *A Primer on Electric Vehicles in India*,
https://doi.org/10.1007/978-981-97-8966-5_10

10.1.2.1 Understanding EVs

Chapter 1: Introduction sets the tone for the book and motivates it. It briefly discusses the statistics on EVs in India in terms of assembly and manufacturing, and uptake. In order to understand the electric vehicle, first, it must be noted that the vehicle has fewer moving parts than the ICE vehicle. As we discuss in Chap. 4: Value Chain for Manufacture and Assembly, the value chain of the EV for manufacture or assembly comprises several steps. Certain parts are common to both ICE and EV: e.g. clutch/torque control, start/stop system, pistons, spark plugs, and turbochargers. The new parts include a power distribution module, onboard charger, battery management system (BMS), vehicle interface control module, DC/DC converter, etc. The most critical part is the battery. Every battery has two electrodes —a cathode and an anode— the cathode largely determines a battery's performance, cost, and environmental footprint. Cathode today is produced almost entirely in Asia.

Second, as we note in Chap. 3 Technology in EVs, the nature of the technology of the batteries determines the material that is used. The battery of the EV uses various minerals like lithium and cobalt. As we discuss in the chapter, lithium is used in lithium-ion batteries predominantly. In EVs, the battery accounts for 70% of the cost of two-wheelers and 50% of cars. The scarcity of minerals like lithium is prompting research into alternatives. Some of these issues have been discussed in the chapter. As R&D is going on worldwide (partly to offset the scarcity of lithium), some of these technologies will become commercial and will determine the demand for critical minerals in the future.

Chapter 7: Policies and Regulations at the central level and States also offers a discussion on the ecosystem of the EV sector. Globally, all big players take around 5–7 years to stabilize production processes and invest heavily in cell R&D. In the Indian case, it must be noted that BMS is not included in the package and is left to the manufacturer to figure out the technology and manufacturing. BMS is critical to ensure battery pack parameters, safety, and performance. Hence a timeline of starting manufacturing within 2 years and achieving 60% localization in 5 years could be a challenge.

10.1.2.2 Supply Chain Challenges in the EV Sector

The main challenges in manufacturing batteries are the availability of critical minerals like lithium, cobalt, and nickel among others. This has been discussed in Chap. 2: Sourcing Minerals for the Battery. All EV production depends on the supply of cobalt and lithium from a few countries. Where are they imported from? Are there alternative sources for these materials? 65% of lithium reserves are in Bolivia and Chile, while 60% of cobalt reserves are in Congo. For all countries, securing a supply of essential metals is critical. For a country like India, securing these critical minerals is essential in fulfilling its EV ambition.

India's strategy in technology acquisition and sourcing minerals to produce low-cost EVs is of paramount importance. As we discuss in Chap. 5: Imports and Exports of Components, certain components are imported. Over the period 2018–19 to 2021–22, electric motors, generators, parts of transformers, etc. are imported. How can India develop capability in these components? Unless capability is developed in the sector, this has implications for India's EV transition. One of the important lessons from the supply shocks of Covid-19 is the need for resilience in value chains. There are geopolitical [1] implications of the above as all countries grapple with sourcing minerals.

10.1.2.3 Opportunities for Indian Companies

The EV transition will present opportunities for investment, jobs, and exports in the future. Chapter 6: FDI and Investment gap in the EV sector discusses the nature and volume of FDI flowing into the sector. Investments made by various companies have also been discussed in the chapter. FDI brings along technology with it. Is technology transfer also taking place? Will this help India develop its capability in the sector? What can India do to attract more investment?

Exports of certain components are already taking place—these include Motor vehicles and motor vehicle components, parts, accessories, etc., static converters, etc. This was discussed in Chap. 5: Imports and Exports of Components. Along with localization due to the PLI schemes, Chap. 7: Policies and Regulations at the central level and States also offers a discussion on how the EV sector has been regulated and it has contributed to the Make in India and Phased Manufacturing Programme of the Government of India.

The transition from ICE to EVs will require changes in the workforce. As we discuss in Chap. 9: Skills and workforce training, certain skills are required in ICE which may become redundant in EVs. The chapter also discusses the skills required in EV assembly and manufacturing, and repair. It will examine the training necessary for inculcating this skill set in workers. What the EV transition will mean for the country in terms of the workforce engaged in ICE versus EVs, has also been discussed in Chap. 9. The EV transition also presents an opportunity to upskill the workforce.

10.1.2.4 Other Themes

Through this book, we hope to answer some questions such as where is India in the EV transition. What are the major challenges in the EV ecosystem? What is working in the regulatory space? The book concludes with some thoughts on the way forward by bringing geopolitics into the discussion. Some of the other themes that have been discussed include:

Financing EV Transition in India.

Financing the EV transition will be key. In Chap. 8: Financing EV Transition in India, we discuss if private domestic investment is flowing into the sector.

Industrial policy and EV transition.

The role of industrial policy in promoting national champion firms is evident in Chap. 7 which discusses the PLI as well as the FAME I, and FAME II schemes.

Value Chain of EVs: Upstream

Part II: Upstream

Part II contains Chaps. 2 and 3. These chapters discuss the upstream value chain of the battery chain and focus on the minerals and technology used to produce EV batteries.

Downstream

Part III: Making in India

Section III of the book deals with Manufacturing and Assembling in India in the EV value chain. This part has three chapters. Chapter 4 is on the value chain. Chapter 5 is on the imports of components like batteries that are not manufactured in the country currently but have export potential in the future when the capability is built. Chapter 6 discusses the future of the industry in terms of the realization of the investment currently made. Chapter 7 discusses the policy landscape of the country.

Ecosystem

Part IV: Ecosystem

Part IV will discuss the evolution of the ecosystem in India. It will discuss the conditions that lead to the development of an ecosystem and how an ecosystem evolves thereafter. It will also discuss the role of industrial and trade policy in this context. It will discuss these issues in Chap. 7 on policies and regulations in the EV space. Chapter 8 is on financing the EV transition in India. Finally, Chap. 9 discusses skills and employment in the EV industry.

At the centre, there are policies like FAME I and FAME II. PLI scheme for Advanced Cell Chemistry has also been announced. In addition, most states have an EV policy. Are these policies and regulations enough to usher in an EV transition? Is the PLI scheme adequate to address this challenge to develop capability? However, since none of the players allocated PLI subsidy (including Reliance) have Li-Ion cell technology at present, it would be interesting to see how it all pans out for all new players with completely new technology, know-how, and processes. These are some of the issues that Chap. 7 addresses.

Electric mobility [2].

The Indian government has a desire to achieve a 100% switch to clean energy for all vehicles in the country by 2030. Along with contributing to climate change targets, it is estimated that a 100% switch to electric vehicles would cut air pollution in Indian cities by 80–90%, bring down India's oil imports by 10% (imported oil currently meets 98% of its energy requirements) and open up opportunities for investment in a new sector. However, this target was later scaled down to 30% electric vehicles by 2030, as a more realistic goal. The Indian Government has been undertaking various measures to boost electric car sales in the country.

Geopolitics

Jones et al. (2023) say that as the demand for electric vehicles is increasing, it is also shaping the geopolitical landscape. The move away from fossil fuels to critical minerals for the EV supply chain is in large part responsible for the change. According to Capri (2021), techno-nationalism is being spawned through the sector, reminiscent of neo-mercantilism due to the use of subsidies in many countries.

Way Ahead

Part V: Looking Ahead

The final part of the book examines the way ahead, the geopolitics behind mineral sourcing, and the path for EV transition. Chapter 10: Conclusion and way ahead brings together all the strands explored in the book by way of a conclusion. There are some possible pathways for the EV transition in India. The pathways will depend on a host of factors including geopolitics.

10.1.3 Gaps in the Discussion

In the book, while most issues pertaining to the EV transition have been covered, there are some issues that have not been addressed. For e.g. Electricity generation from renewable sources has not been covered. This is critical if the EV transition is to really help in cutting down emissions. The other issue that has not been covered is Air pollution. While this is an important issue and there have been estimates of how much emissions reduction could result as a result of the EV transition. However, addressing this question adequately requires a life cycle assessment of the EVs and ICEVs. This is beyond the scope of the present volume. Also, the role of cities in promoting electric mobility has not been discussed. Also, the discussion has not referred to what consumers want [3] or from the consumer perspective. The issue of the safety of electric vehicles has very briefly been alluded to.

10.1.4 Way Ahead

There are some possible pathways for EV transition which partly depend on geopolitics. First, if the country progresses on the path it has embarked on the EV transition will certainly lead to an increase in the vehicles bought, assembled, and made in the country. Second, however, another possible pathway is that India decarbonizes its energy production, achieving its NDC commitment and that makes the EV transition also fully green. There are also numerous possible pathways in between, depending on the pace of the decarbonization process. The role of geopolitics comes from the access to minerals, materials, technology, and finance in achieving the best outcome possible.

Machine Generated

Morfeldt and Johansson (2022) study statistics on Swedish passenger cars between 2014 and 2018, and find that carbon footprint may be substantially reduced if the relationship between average annual driving intensity and vehicle lifetime is weakened. This points to the importance of designing future BEVs (both autonomous and regular) for durability. Nunes et al. (2022) estimate the aggregate utilization (miles travelled over the vehicle's lifetime) and longevity (measured in years of ownership) required for EVs to reduce emissions relative to each counterfactual, considering vehicle status and existing utilization trends. Ojha and Agarwal (2023) simulate 5 scenarios on the trends of market demand, metal quantity extracted, battery lifespan, recollection efficiency, and yield from reprocessing routes. They conclude that the success of the circular economy in the EVB supply chain significantly depends on its EOL period and recollection efficiency. Dixit and Singh (2022) study the decision of consumers to buy an electric vehicle using a predictive machine learning model. was developed to predict whether a consumer in India will or will not buy an electric vehicle depending on the input variables. The model included age, gender, income, level of environmental concerns, vehicle cost, running cost, vehicle performance, driving range, and mass behavior as significant predictors to classify a consumer in either category. Janmala et al. (2021) observe that the EV load penetration has resulted in degrading the feeder performance significantly in terms of increased real power losses, decreased voltage profile, and consequently resulted in lowering the voltage stability margin. They suggest that optimal placement and sizing of DGs and EVs could improve feeder performance. Singhal and Patel (2022) study the barriers to the adoption of e-mobility in India and find that socio-economic barriers are the most important, followed by infrastructural barriers and technological barriers. Arora et al. (2022) study the factors that determine the purchase intention of consumers with respect to EVs in the Indian context. The results indicate that attitude (as consumers believe EVs to be complex in terms of their usability), limited availability, and the absence of required public charging infrastructure. Ilango et al. (2022) propose an efficient hybrid technique for the optimal allocation of electric vehicle charging spots in the distribution network system. The proposed technique is used to evaluate the reliability of the smart grid based on the state matrix incorporation of random parameters of the electric vehicle. Mondal et al. (2023) argue that while a single EV has a negligible impact on the system, the combined effect of charging an EV cluster may violate the voltage security constraints of the RDS. They propose an algorithm that is efficient in maintaining the node voltages during the EV charging process.

References

Capri, A. (2021) The geopolitics of electric vehicles: techno-nationalism reshapes the automotive industry. Hinrich Foundation Report. Link: https://www.hinrichfoundation.com/research/wp/tech/geopolitics-of-electric-vehicles/

Gvosdev, N. K. (2023) The Ethics and Geopolitics of the Electric Vehicle Transition. Carnegie Council for Ethics in International Affairs, Podcast. April 10. Link: https://www.carnegiecouncil.org/media/article/ethics-geopolitics-electic-vehicle-transition

Jones, B., Nguyen-Tien, V., Elliott, R., and Harper, G. (2023) The supply chain for electric car batteries is changing the world's geopolitics. LSE blogs. June 27. Link: https://blogs.lse.ac.uk/businessreview/2023/06/27/the-supply-chain-for-electric-car-batteries-is-changing-the-worlds-geopolitics/

Mckinsey & Company (2023) What EV buyers really want? August 5. Link: https://www.mckinsey.com/featured-insights/themes/what-ev-buyers-really-want

Footnotes/Endnotes

[1] Gvosdev (2023).

[2] Mckinsey (2022).

[3] McKinsey (2023).

10.2 Machine Generated Summaries

Disclaimer: The summaries in this chapter were generated from Springer Nature publications using extractive AI auto-summarization: An extraction-based summarizer aims to identify the most important sentences of a text using an algorithm and uses those original sentences to create the auto-summary (unlike generative AI). As the constituted sentences are machine selected, they may not fully reflect the body of the work, so we strongly advise that the original content is read and cited. The auto generated summaries were curated by the editor to meet Springer Nature publication standards. To cite this content, please refer to the original papers.

Machine generated keywords: allocation, behaviour, distribution network, optimal, voltage, algorithm, pathway, product, bus, carbon footprint, travel, distribution, consumer, carbon, network.

10.2.1 *Impacts of Shared Mobility on Vehicle Lifetimes and on the Carbon Footprint of Electric Vehicles [1]*

This is a machine-generated summary of:

Morfeldt, Johannes; Johansson, Daniel J. A.: Impacts of shared mobility on vehicle lifetimes and on the carbon footprint of electric vehicles [1]

Published in: Nature Communications (2022).

Link to original: https://doi.org/10.1038/s41467-022-33666-2.

If you want to cite the papers, please refer to the original.

For technical reasons we could not place the page where the original quote is coming from.

Abstract-Summary

"Shared cars will likely have larger annual vehicle driving distances than individually owned cars."

"We show that the carbon footprint depends on the cumulative driving distance, which depends on both driving intensity and calendar aging."

"Higher driving intensities generally result in lower carbon footprints due to increased cumulative driving distance over the vehicle's lifetime."

"Shared cars could decrease the carbon footprint by about 41% in 2050, if one shared vehicle replaces ten individually owned vehicles."

Introduction

"The cumulative driving distance over the vehicle's lifetime is an important assumption when estimating the carbon footprint of passenger car travel but varies significantly among studies [2] and has been shown to have large impacts on the results [3]."

"Considering a relationship between driving intensity and vehicle lifetime is critical when assessing the carbon footprint of shared autonomous BEVs."

"The carbon footprint-related consequences of changes in driving intensity in response to shared autonomous BEVs and plausible levels of empty travel have not yet been analyzed in situations where energy systems are decarbonized over time."

"The vehicle's lifetime, its annual driving intensity, and its interaction with decarbonizing energy systems will play important roles for the carbon footprint of passenger car travel over the coming decades."

"We aim to bridge this research gap by estimating the impact of vehicle lifetime and annual driving intensity on the carbon footprints of passenger cars used for sharing."

Results

"The remainder of this article explores how changes in annual driving intensity may influence the carbon footprint of passenger car travel, assuming that the relationship shown for ICEVs is applicable as a proxy for individually owned and shared autonomous BEVs."

"We capture the uncertainty in future vehicle lifetimes of (shared and autonomous) BEVs by highlighting extreme values for the relationship between annual driving intensity and vehicles lifetime as well as the empirically estimated relationship based on ICEV retirement data."

"$\varepsilon = 0$ is a relevant extreme case if future individually owned and/or shared autonomous BEVs would age in a way where driving intensity has no importance in the decision to retire vehicles."

"Though that the combination of a shared autonomous BEV replacing ten individually owned BEVs and assuming 100% empty travel results in high annual driving intensity of ca 280,000 km, which may not be possible to achieve for one car."

Discussion

"Our analysis shows that the relationship between vehicle lifetime and driving intensity is an important factor when estimating the carbon footprint of shared mobility."

"Increasing the driving intensity, for example through shared autonomous BEVs, may be an option for reducing lifecycle emissions from passenger car travel."

"The results indicate that shared autonomous BEVs would reduce the carbon footprint if it results in higher driving intensity of each individual vehicle."

"We find that a system with shared autonomous BEVs can decrease the carbon footprint per kilometer of intended travel by about 41% if one shared vehicle replaces 10 individually owned vehicles in 2050."

"The analysis shows that the carbon footprint may be substantially reduced if the relationship between average annual driving intensity and vehicle lifetime is weakened, pointing to the importance of designing future BEVs (both autonomous and regular) for durability."

Methods

"Statistics on Swedish passenger cars retired between 2014 and 2018 are used to understand how changes in annual average driving intensity could influence vehicle lifetimes."

"The semi-empirical lifetime-intensity model enables estimations of vehicle lifetime probabilities for a given annual average driving intensity."

"Use-phase CO_2 emissions, E_{Use}, are estimated annually based on total traveled distance (including potential empty travel), vehicle energy use, and appropriate carbon intensities described below."

"The carbon intensity of electricity is based on estimates of average direct emissions for future electricity mixes for each respective geographic area, see description of the data sources for the scenarios below."

"For the global electricity mix used in manufacturing and for charging, future direct emissions and adjustments to account for transmission and distribution losses (based on the difference between estimated supply and demand) are based on estimates by the IEA [4] for the two decarbonization pathways, Stated Policies and Sustainable Development."

10.2.2 Re-thinking Procurement Incentives for Electric Vehicles to Achieve Net-Zero Emissions [5]

This is a machine-generated summary of:

Nunes, Ashley; Woodley, Lucas; Rossetti, Philip: Re-thinking procurement incentives for electric vehicles to achieve net-zero emissions [5].

Published in: Nature Sustainability (2022).

Link to original: https://doi.org/10.1038/s41893-022-00862-3.

If you want to cite the papers, please refer to the original.

For technical reasons we could not place the page where the original quote is coming from.

Abstract-Summary

"Their effectiveness in reducing transportation emissions depends on the behavioural characteristics of EV adopters."

"When an EV is used, under what conditions and by whom dictates whether or not these vehicles can deliver emissions reductions."

"We document that replacing gasoline powered vehicles with EVs may—depending on behavioural characteristics—increase, not decrease, emissions."

"We conclude that achieving emissions reductions using EVs requires redesigning procurement incentive programmes in a manner that (re)distributes incentives towards the second-hand EV market."

Main

"Can existing electric vehicle (EV) subsidy policies facilitate reductions in carbon emissions?".

"Less clear, however, is the extent to which an EV's emissions advantage persists given heterogeneity in consumer behaviour."

"We assess how diversity in consumer behaviour impacts an EV's ability to deliver an emissions advantage."

"Whereas past studies have examined EV's potential emissions advantages given heterogeneity in behavioural patterns [6, 7] or counterfactual vehicle fleets in isolation [8], the present study examines how the confluence of driving behaviour and choice of vehicle replacement impact EV's emissions benefits."

"By jointly analysing behavioural heterogeneity and counterfactual vehicle inventory (which vehicles a household would have owned or purchased were an EV not procured), our study demonstrates how the preconditions necessary to deliver an EV emissions advantage vary across households."

"Whereas other studies assume static and homogenous consumer behaviour as a precursor to assessing (and realizing) EV's emissions advantage [8], we quantify requisite EV mileage needed to achieve an emissions advantage given documented behavioural heterogeneity in how vehicles are used in multi-vehicle households [6, 7]."

Results and Discussion

"We estimate the aggregate utilization (miles travelled over the vehicle's lifetime) and longevity (measured in years of ownership) required for EVs to reduce emissions relative to each counterfactual, considering vehicle status and existing utilization trends."

"Realizing emissions benefits from EVs instead probably requires—our analysis suggests—reducing emissions from both fuel production and vehicle manufacturing, as simultaneous improvements to both factors generate highly elastic reductions to the requisite longevity and aggregate utilization thresholds of EVs."

"We argue that the existing structure of EV subsidies, which predominantly favours new vehicle adoption, is not only less likely to deliver envisioned emission benefits than policies that incentivize longer EV retention but may also produce regressive results, directing financial relief away from households that are (1) more likely to need it and (2) more likely to facilitate realization of an EV's emissions advantage."

Methods

"EVs refers to vehicles solely powered by electricity obtained from the power-generating electric grid (estimated to generate 436 $gCO_2e\ kWh^{-1}$, ref [9]); counterfactual vehicle inventory refers to the number of vehicles a household would own were an EV not procured; utilization refers to EV mileage (considered on aggregate and annualized bases); and longevity refers to how long an EV remains in service (expressed in years)."

"We analyse responses regarding household vehicle ownership magnitude (how many vehicles a household has) and the total vehicle miles travelled by those vehicles annually."

"We subsequently estimate (1) the proportion of households nationwide that have zero, one, two, three and four or more vehicles in household inventory and (2) their contribution (expressed as a percentage) to total vehicle miles (VMT) travelled."

"Regarding the procurement of EVs, we assume that EV procurement incentives alone do not induce a household to purchase a greater number of vehicles than it otherwise would."

10.2.3 Implications of Circular Production and Consumption of Electric Vehicle Batteries on Resource Sustainability: A system Dynamics Perspective [10]

This is a machine-generated summary of:

Ojha, Ravindra; Agarwal, Alpana: Implications of circular production and consumption of electric vehicle batteries on resource sustainability: A system dynamics perspective [10].

Published in: Environment, Development and Sustainability (2023).

Link to original: https://doi.org/10.1007/s10668-023-03279-w.

If you want to cite the papers, please refer to the original.

For technical reasons we could not place the page where the original quote is coming from.

Abstract-Summary

"Circular economy in the electric vehicle battery (EVB) ecosystem can provide a sustainable solution to resource management."

"This has triggered the need to understand and analyse the EVB ecosystem players."

"Long-term analyses of behaviours of the different variables under dynamic and complex systems are best carried out by the application of system dynamics (SD)—a continuous simulation methodology."

"This research paper has contributed in developing an interactive SD-based model for the circular production and consumption flow of EVBs to analyse the behavioural dynamics of the different variables involved."

"The behaviour trends of the variables obtained from the SD-based simulation model have provided useful insights and five recommendations for resource sustainability."

Introduction

"Given the growing EV market, customer expectations of affordability and more efficient products call for technological advancement."

"The EVB manufacturing processes rely more on a linear supply chain, wherein raw materials are extracted from the environment; new products are produced and disposed after use, leading to the accumulation of waste."

"There is a compelling need to develop recycling and remanufacturing technology (Zhao and others [11]) or adopt the circular economy (CE) principle."

"There is a compelling need for comprehensive research to understand, explore and analyse the dynamics of the variables in the EVB ecosystem."

"This can be carried out using the system dynamics (SD) simulation approach in the current I4.0 smart manufacturing environment, which has the potential for CE and sustainability."

"This research paper aims to develop an SD-based model for the circular production and consumption flow of EVB to analyse the behavioural dynamics of the different variables involved."

Literature Review

"With I4.0, companies can efficiently manufacture quality products with optimal resource consumption."

"I4.0 is also capable of creating smart and interconnected production processes for EV manufacturing."

"Lithium-ion batteries, the most important aggregate in EVs, have become more relevant for the economy and environment (Sun and others [12]) for fulfilling the raw material requirement during production."

"The use of lithium-ion batteries in manufacturing electric vehicles will need a high amount of natural resources such as nickel, cobalt, lithium, and copper (Alessia and others [13]; Aichberger & Jungmeier [14])."

"This is an opportunity for EV manufacturers also to repair, refurbish, and remanufacture the batteries (Gautam and others [15]; Guarnieri and others [16]; Jabbour and others [17])."

"The promising benefits of the current trends of I4.0, the expected adoption trend of EVs to an unprecedented scale, the operational challenges of the EVB ecosystem, and environmental sustainability (Wendler [18]) have together motivated the authors to understand the dynamics in it."

Research Methodology

"Metal required for production = EV market demand × metal used in each EV Collection efficiency for reprocessing = Fraction (assumption) × EVBs in the system

Dump = unrecoverable material = (1 − collection efficiency) EOL material 'Delay3' function of SD approach applied as DELAY3(< input > , < delay duration > [, < initial >]) (3) Assumption considered in the Stella software application: A realistic and commonly applied third-order delay (DELAY3) function SD was considered for the following variables: extraction of mineral, battery EOL span, and recirculation of EOL battery through the four recirculation routes with different delay periods."

"A comparative behaviour trend of net virgin metal (NVM) was studied under the following two conditions: (i) varying recollection efficiencies (20, 40 and 75 per cent) but with the same demand growth rates and EOL of EVB and (ii) varying EOL (5, 8 and 10 years) of EVB but with same demand growth rates and recollection efficiency."

Results and Discussions

"In the period thereafter, the recirculated metal quantity increases due to high recollection efficiency (60%), and therefore, the NVM quantity needed starts reducing briskly."

"The reduced NVM quantity, in comparison with scenario S-2, can be assigned to lower battery life (84 months) leading to the quicker entry of the EOL battery into the recirculation cycle and a better recollection efficiency trend."

"The area under the curve is observed to be increasing with improving battery life leading to additional mining of minerals and thereby creating additional pressure on natural resources."

"With improved battery life, the EOL batteries entering into the recirculation cycle get delayed by the differential EOL period, leading to a lower quantity collected for circulation, over time, in an already growing EV market."

"A simulation run with a significantly high (80%) recollection efficiency, battery life of 10 years, and the same demand trend has shown a very favourable trend with a significantly low requirement of NVM."

Recommendations

"With the current trends of EV market demand, growing technology, and mass customization, pressures of environmental sustainability, responsiveness and affordability expectations of customers, and improving process digitalization, embracing I4.0 has become a necessity in the EV ecosystem."

"With lithium-ion batteries being the critical aggregate in the EV vehicle, circular flow in the supply chain (Rosendahl & Rubiano [19]) has gained momentum for environmental sustainability."

"Recollection efficiency of the EOL EVB has a significant impact in terms of virgin metal requirement."

"Higher recollection value and its improvement trend would significantly ease the pressure to extract metals from limited mineral resources, leading to better environmental sustainability."

"A significantly attractive incentive system by policymakers can play a booster role in improving the recollection efficiency of EVBs after their EOL."

"Technology coupled with big data analytics of I4.0 can help in optimizing battery life and recollection efficiency."

Research Implications

"This paper has practical and theoretical implications for environmental and business researchers and practitioners."

"Of the simulation of five scenarios built around I4.0 cum CE-enabled EVB manufacturing processes, several recommendations made are: strategic I4.0 adoption, CE awareness, and attractive incentive system to encourage recollection."

"Besides, the scenarios built by rationally mapping EVB supply chain variables, CLD/SFD, and SD model created in this paper can have high implications."

"These can be further studied by researchers and practitioners to understand the systems behaviour."

Conclusions

"With the current advancing trends of digitalization, burgeoning demand for EVs, and the dire need for circular production–consumption of EVBs, the technology-enabled I4.0 will play a critical role in the sustainability of the battery ecosystem."

"The five scenarios simulated have provided useful insights through the trends of the following key variables: market demand, metal quantity extracted, battery lifespan, recollection efficiency, and yield from reprocessing routes."

"The success of the circular economy in the EVB supply chain significantly depends on its EOL period and recollection efficiency."

"Manufacturers, scientists, supply chain managers, and industry leaders are working to innovate the recollection methods, improve reprocessing systems and enhance the product life cycle of EVBs."

Research Limitations and Scope for Further Research

"This research paper has a few limitations and there is scope for future research work."

"The financial dimensions of the EVB business have not been considered in the model and therefore, is a limitation to work in the future."

"Three other proposed areas for further research are (i) a study on the potential implications of I4.0 on EVB ecosystem, (ii) strategies for incentivizing circular economy in EVBs and analysis after that, and (iii) SD modelling and analysis of the economic viability of the CE of EVB ecosystem."

10.2.4 Predicting Electric Vehicle (EV) Buyers in India: A Machine Learning Approach [20]

This is a machine-generated summary of:

Dixit, Sushil Kumar; Singh, Ashirwad Kumar: Predicting Electric Vehicle (EV) Buyers in India: A Machine Learning Approach [20].

Published in: The Review of Socionetwork Strategies (2022).

Link to original: https://doi.org/10.1007/s12626-022-00109-9.

If you want to cite the papers, please refer to the original.

For technical reasons we could not place the page where the original quote is coming from.

Abstract-Summary

"With advancements in technology, electric vehicles (EVs) have shown a new potential to meet many of the challenges being faced by humanity."

"The adoption of electric vehicles has remained challenging despite consumers having a positive attitude toward EVs and big policy pushes by governments in many countries."

"Marketers from the electric vehicle (EV) industry are finding it difficult to identify genuine buyers for their products."

"The present study attempts to develop a machine learning model to predict whether a person would "Buy" or "Won't Buy" an electric vehicle in India."

"To develop the model, an exploration of EV context was done first by conducting a text analysis of online content relating to electric vehicles."

"The machine learning model indicates that age, gender, income, level of environmental concerns, vehicle cost, running cost, vehicle performance, driving range, and mass behavior are significant predictors of electrical vehicle purchase in India."

Introduction

"A study focusing on consumer concerns for electric vehicles and understanding the factors affecting electric vehicle uptake was missing in the Indian context."

"The text analysis of social media posts along with the review of existing literature provides insights into the context of the Indian electric vehicle industry, as well as the possible factors affecting EV uptake in India."

"Kumar and others [21] studied challenges to the adoption of electric vehicles and concluded that sharing economy and public utilities will play a critical role in EV purchase, considering the high cost coupled with consumers' low purchasing power and lack of availability of charging infrastructure in India."

"There is no comprehensive list of the factors affecting electric vehicle uptake; thus, the present study aims to not only list the factors that are relevant in the current Indian context, but also develop a model to predict who will buy an electric vehicle in India."

Analysis and Findings

"A word cloud was formed for better visualizations, which also became the basis for the questionnaire that was later floated to make a predictive model that could classify whether a consumer will "Buy" or "Won't Buy" an electric vehicle."

"A predictive machine learning model was developed to predict whether a consumer in India will or will not buy an electric vehicle depending on the input variables."

"The predictive machine learning model developed can classify whether an Indian consumer would "Buy" or "Won't Buy" an electric vehicle."

"The constant of the model was also reported to not be a statistically significant predictor of electrical vehicle purchase."

"The model included age, gender, income, level of environmental concerns, vehicle cost, running cost, vehicle performance, driving range, and mass behavior as significant predictors to classify a consumer in either category."

10.2.5 Future Search Algorithm for Optimal Integration of Distributed Generation and Electric Vehicle Fleets in Radial Distribution Networks Considering Techno-environmental Aspects [22]

This is a machine-generated summary of:

Janamala, Varaprasad; Kamal Kumar, U.; Pandraju, Thandava Krishna Sai: Future search algorithm for optimal integration of distributed generation and electric vehicle fleets in radial distribution networks considering techno-environmental aspects [22].

Published in: SN Applied Sciences (2021).

Link to original: https://doi.org/10.1007/s42452-021-04466-y.

If you want to cite the papers, please refer to the original.

For technical reasons we could not place the page where the original quote is coming from.

Abstract-Summary

"A new nature-inspire meta-heuristic algorithm called future search algorithm (FSA) is proposed for the first time to solve the simultaneous optimal allocation of distribution generation (DG) and electric vehicle (EV) fleets considering techno-environmental aspects in the operation and control of radial distribution networks (RDN)."

"A techno-environmental multi-objective function is formulated using real power loss, voltage stability index."

"The active and reactive power compensation limits and different operational constraints of RDN are considered while minimizing the proposed objective function."

"The effectiveness of the proposed methodology is presented using different case studies on Indian practical 106-bus agriculture feeder for DGs and 36-bus rural residential feeder for simultaneous allocation of DGs and EV fleets."

Introduction

"In [23], bacterial foraging optimization algorithm (BFOA) is proposed for simultaneous allocation of DGs and CBs towards loss minimization and voltage profile improvement."

"In [24], weight improved particle swarm optimization algorithm (WIPSO) and self adaptive differential evolution algorithm (SADE) have been proposed for solving the simultaneous allocation of renewable type DGs and CBs by minimizing the multi-objective function formed using real and reactive power losses and voltage profile."

"In [25], ant lion optimization (ALO) algorithm is proposed for minimizing the multi-objective function using real power loss, voltage profile and voltage stability via optimally locating and sizing the renewable DGs."

"The optimal allocation of DGs problem is addressed using quasi-oppositional chaotic symbiotic organisms search (QOCSOS) algorithm by aiming for reduction in real power loss, improvement in voltage profile and enhancement in voltage stability [26, 27]."

"Coyote optimization algorithm (COA) [28] is proposed for optimal integration of DGs in RDNs considering multi-objective function, formed using real power loss, operating cost and voltage stability."

Modeling of Relevant Concepts

"As per the active and/or reactive power compensation, these DGs can be modeled as negative loads and their classification along with modeling as follows: Photovoltaic, micro turbines and fuel cells which are incorporated into the main grid with the help of converters/inverters are good examples of real power compensation sources."

"The real power load at bus-i will be reduced by an amount equal to DG's real power output with unity power factor (upf) and is given by: Shunt CBs (in either fixed or/ and switched form) are the best examples for reactive power compensation sources particularly at the distribution network."

"The reactive power load at bus-i will be reduced by an amount equal to DG reactive power output and is given by: Wind farms and synchronous generators are example of complex power compensation sources."

Problem Formulation

"This can be achieved mainly by improving the voltage profile across the network, and consequently loss can reduce by having reduced current flow through each branch/ element."

"reference voltage (V_r) has to be minimized for having adequate voltage profile and stability margins and consequently reduced real power losses in the network can be achieved significantly."

"By various power system blackout incidents across the world, voltage stability has become one of the primary concerns of the system operator [29]."

"By optimizing the penetration level of different types of DGs and EV fleet, the network voltage stability is aimed to maximize."

"By reducing the real power loss and dependency on grid power via optimal allocation of DGs and EV fleets, the power generation from grid reduces and consequently environmental benefits can be achieved."

Future Search Algorithm

"The others HSA start its means by a random population and it formulates its iterations dependent on the best optima of the random beginning."

"This best population might be a long way from the initial population, this makes the HSA take many iterations to come to the global optima."

"In every HSA, there is a local best population between the multi-variables and global optima between the iterations."

"The FSA uses the local best solution and the global optima for finding the best population."

"The objective function of each initial population is defined as a local solution (LS) and the best among all is defined as a global solution (GS)."

"Defining new GS and LS at each iteration and again using them for updating the optimal solution within the same iteration is the key feature of FSA for outperformance than the other HSA."

Results and Discussion

"Case 3: Feeder performance improvement via optimal allocation of DGs with upf considering loss minimization and VSI improvement: By using proposed FSA for solving DGs allocation problem at this case, the best solution obtained is as follows: the locations of DGs are 36, 9 and 35 and their sizes in kW are 550, 541 and 1073 respectively."

"Case 4: Feeder performance improvement via optimal allocation of EV fleet considering loss minimization: In this case, the increased 10% of loading condition (279.11 kW + j 192.29 kVAr) across the network is taken as a load of single EV fleet and optimized its location."

"Case 5: Feeder performance improvement via simultaneous optimal allocation of DGs and EV fleet considering loss minimization and voltage stability maximization: In this case, the increased loading condition across the network is taken as total power rating of a single EV fleet and optimized its location along with DG locations and sizes."

Conclusion

"The simultaneous allocation of DGs and EVs has been formulated as a multi-objective function using real power loss, average voltage deviation index (AVDI), voltage stability index (VSI) and greenhouse gas (GHG) emission."

"From the case studies, the EV load penetration has resulted in degrading the feeder performance significantly in terms of increased real power losses, decreased voltage profile and consequently resulted in lowering the voltage stability margin."

"In counter to this scenario, the optimal placement and sizing of DGs and EVs resulted for improvement in feeder performance significantly."

"Validation of proposed methodology for enhancing the performance of distribution network by using simultaneous smart charging scenario in EV fleets and timely optimal network reconfiguration w.r.t."

"The computational efficiency of the proposed meta-heuristic FSA algorithm in solving such complex multi-objective optimization problems needs to be further investigated and treated as future scope of this research."

10.2.6 Prioritisation of Barriers in the Implementation of Electric Mobility in Indian Context Using Fuzzy Analytical Hierarchical Process [30]

This is a machine-generated summary of:

Singhal, Ruchi; Patel, Chetan R.: Prioritisation of Barriers in the Implementation of Electric Mobility in Indian Context Using Fuzzy Analytical Hierarchical Process [30].

Published in: Journal of The Institution of Engineers (India): Series A (2022).

Link to original: https://doi.org/10.1007/s40030-022-00676-8.

If you want to cite the papers, please refer to the original.

For technical reasons we could not place the page where the original quote is coming from.

Abstract-Summary

"This study aims to identify and prioritise the barriers in the adoption of e-mobility."

"Results demonstrated that socio-economic barriers got the maximum weight, followed by infrastructural barriers and technological barriers."

"It reveals that if negative perception towards the electric vehicle is not improved, then it would be challenging to make e-mobility a success story."

Introduction

"High growth in penetration rate of 3 W has been observed without much government intervention, penetration in 2 W and 4 W has always been low which shows that the concept of electric mobility is in the emerging stage and has some barriers which are affecting the EV adaptation in developing countries like India."

"For the study, FAHP is used to rank the identified barriers in the acceptance of electric vehicle using expert opinion."

"The objectives of the study were (1) to discover the factors influencing the adoption of e-mobility in the Indian context; (2) to select the most suitable method for this type of multi-criteria decision-making; and (3) to analyse the priority barriers considering the opinion of experts in this field."

"The study will help the researchers and professionals understand factors acting as barriers to adoption of e-mobility, especially in the developing economies like India, where adoption is way less than targeted."

Research Framework

"Comprehensive literature review of studies related to the subject was done to list the barriers in the adoption of e-mobility."

"The questionnaire was prepared to allow the pairwise comparison of barriers using five-point scales."

"The domain experts who have exposure in policy preparation, urban transport studies, social perception and other parameters concerning implementation and use of electric mobility were invited to opine on barriers."

"According to the level of the criticalness of barriers, their prioritisation has been done."

Literature Background

"Mainly, barriers in the implementation of e-mobility have been identified by reviewing worldwide literature, and then the analysis has been done in the Indian context using multi-criteria decision-making (MCDM) method."

"These are as follows: (1) literature related to barriers in the adoption of e-mobility and (2) literature on the latest MCDM methods, used to deal with such problems."

"For finding the studies related to latest MCDM methods, the articles from 2012 onwards were searched in the database with a keyword, 'multi-criteria decision-making', through which 58,114 studies appeared."

"The frequently used MCDM methods for such problems are analytical hierarchical process (AHP), Decision Making Trial and Evaluation Laboratory (DEMATEL), Preference Ranking Organisation Method for Enrichment Evaluation (PROMETHEE) and the Technique for Order of Preference by Similarity to Ideal Solution (TOPSIS)."

"Chang's FAHP method had been used to solve multi-criteria decision-making problems in various fields, e.g. to evaluate environmental sustainability [31], coastal erosion risk factors [32], procurement performance measurement [33] and decision-making in fuzzy environment [34]."

Evaluation of Weights of Barriers by FAHP

"Chang's method of FAHP has been applied in the following six steps, to rank the barriers in the implementation of e-mobility in an Indian context: Step I Organising the hierarchical chart of the problem."

"The experts were asked to make a comparison using a five-point scale of TFNs to mark the relative importance of barriers and sub-barriers."

"Step V Computing the crisp weight of barriers from their fuzzy synthetic value."

"The degree of possibility is computed by comparing of one synthetic value with other synthetic value denoted by V ($S_1 \geq S_2$) according to the following definition: Assuming, d' (B_i) = min V ($S_i \geq S_k$), and then determining the final weights (W_{Bi}) by normalisation as shown below: Therefore, $W_{\text{Main Barriers}} = W_{Bi} = (0.281, 0.172, 0.365, 0.146, 0.035)^T$, where i = 1, 2, ..., 5 By following the same steps, weights of sub-barriers of each main barrier can be obtained."

Result and Discussion

"It can be seen that socio-economic barriers (B2) got the maximum weight (0.2818) and, therefore, has been ranked as the most critical barriers in acceptance of EVs."

"Two barriers in priority are B21 and B23, respectively, from socio-economic barriers category again out of which B21, i.e. the long charging period has got considerable weightage by all the experts' groups, whereas for B23, i.e. range anxiety, the weightage by decision-makers, service providers and consultants coincides approximately and is higher than B21, but it is pulled down by the rating of academicians."

"In case of service providers who are expert in providing EV infrastructure and mobility solution and services, it can be seen that governance and finance barriers (B4) have got the highest weight (0.2794) as they face lack of support from the government side."

Conclusion

"This study aimed to identify the factors acting as barriers in the transition to e-mobility and prioritise these barriers based on criticalness."

"Ranking of sub-barriers discovered the high initial cost of EVs as most critical sub-barrier in the adoption of e-mobility in Indian context followed by a shortage of charging infrastructure, lack of standard charging infrastructure and poor-tech savvy level."

"The vital contribution of this study is to cope up with e-mobility barriers, to come out with ways to increase electric vehicles penetration in India."

"Prioritisation of barriers in the study will support policymakers in designing more suitable and comprehensive policies to address e-mobility barriers considering all the stakeholders."

"The wide literature survey of e-mobility barriers in this study can be advantageous for scholars working in this field."

"The systematic framework discussed in this study could be a guide for similar studies on e-mobility barriers in other countries although the number of barriers and outcomes may vary."

10.2.7 Using Diffusion of Innovation Framework with Attitudinal Factor to Predict the Future of Mobility in the Indian Market [35]

This is a machine-generated summary of:

Arora, Subhash Chander; Sharma, Megha; Singh, Vinod Kumar: Using diffusion of innovation framework with attitudinal factor to predict the future of mobility in the Indian market [35].

Published in: Environmental Science and Pollution Research (2022).

Link to original: https://doi.org/10.1007/s11356-022-23149-8.

If you want to cite the papers, please refer to the original.

For technical reasons we could not place the page where the original quote is coming from.

Abstract-Summary

"There has been a growing voice from segments of the society for a behavioural shift towards more sustainable modes of transport like usage of electric vehicles (EVs), mobility sharing and rental services and public transportation powered by electricity."

"In India, the statistics or stock of EVs continues to remain poor when compared to its counterparts like China."

"This study is conducted to predict the influence of factors on purchase intention of consumers with respect to EVs in the Indian context."

"The results indicate that attitude has the most influential impact on purchase intention followed by related advantages but consumers also believe EVs to be complex in terms of their usability, limited availability and the absence of required public charging infrastructure."

"The mediating role of attitude between determinants of diffusion and purchase intention was also established except in the case of complexity."

"This study is contemporary and adds to the limited literature available on EV adoption in the Indian market by integrating attitudinal factor within the DOI framework."

"It also provides suitable insights to further the development of EVs in the domestic and global market."

Introduction

"The diffusion has been stagnant in India for a variety of reasons like poor awareness and understanding of the public, scepticism towards EVs and consumers viewing internal combustion engine (ICE) vehicles as more convenient than this new technology (Pradeep and others [36])."

"Taking care of this previous work, the purpose of this study intends to amalgamate both these perspectives and develop a conceptual framework based on diffusion of innovation (DOI) framework along with attitudinal factor to predict the intentions of Indian consumers in adopting this novel technology."

"Stakeholders will get to know which of the five traits of the innovation best predicts the intention of consumers and whether these five aspects on their own can explain the acceptance of EVs or it leads to development of behavioural attitude initially which then becomes the reason for purchase behaviour."

Literature Review and Hypothesis Development

"It can positively influence consumers' EV purchase intentions (Xu and others [37])."

"It is suggested that a higher number of other electric vehicles such as electric passenger cars on roads could lead to development of positive attitude and subsequent purchase intentions among the consumer groups."

"The study by Kahn [38] found positive influence of attitude and perceptions on adoption behaviour of consumers."

"Degirmenci and Breitner [39] found strong influence of environmental performance of these vehicles on the attitude and subsequent purchase intentions of consumers."

"We hypothesise that: H6: Attitude has a positive influence on purchase intention towards EVs."

"Grasping from previous literature, it can also be highlighted that intentions are studied in comparison to actual adoption because in EV's case, the market is still developing and the consumer behaviour is infrequent (Rezvani and others [40]; Sajjad and others [41])."

Research Methodology

"A total of 298 customers spread across the Delhi NCR region were approached on the basis of purposive convenient sampling (Jaiswal and Kant [42]) to collect the required data on variables identified in this research."

"These respondents are aware of EVs, the different models available in the market and the present state of air pollution in Delhi region."

"Each of these variables is indicated by a certain number of measurement items (appended at the end) for which appropriate referencing has also been made."

"Certain items were modified so as to improve the content validity, to bring clarity and that each item is understood in its right spirit (Shalender and Sharma [43])."

"Before collecting the data, pilot testing was done on 50 random respondents to see whether the items are being easily understood by the respondents and were getting interpreted in the right manner so as to take the research in the right direction."

Data Analysis and Results

"This technique shows a theoretical concept in a structural equation and involves a two-step process (Anderson and others [44]) in which we test construct's reliability and validity in stage 1 while in stage 2, we examine the relationships between the given constructs and also determine the overall model fit."

"It is as follows: Guidelines proposed by Hair and others [45] have been taken into account for measuring the reliability and validity of the various constructs."

"Construct validity deals with the accuracy of measurement and is established by two tests: convergent and discriminant validity."

"By convergent validity, we mean that the items of the same construct should converge strongly and is represented by the score of average variance extracted (AVE)."

"To calculate discriminant validity, square root of AVE (bold elements) between two constructs is done and if this value is found to be more than the squared correlation between each pair of latent variables (off-diagonal elements), the criteria is being met."

Discussion and Conclusion

"This study was undertaken to investigate consumers' purchase intentions towards EVs in the Indian market which is characterised by slow diffusion process and poor awareness of the domestic consumers."

"This is understandable as the presence of better features like lower operational costs in EVs exercises considerable influence on buying behaviour and is sufficient ground to spark purchase intention among consumers."

"The study by Li and others [46] indicated that trialability positively changed the attitudes and intentions of consumers once they drove the EVs."

"Better understanding of the consumer-oriented factors can help providers of EVs to initiate policies that better suit the local market conditions and leads to market development in the near future."

"The study investigates consumers' intentions towards EVs in a growing Indian automobile market from the perspective of innovation diffusion theory."

10.2.8 Charging Demand Based on the Interaction Among Electric Vehicles and Renewable Energy Sources Using Hybrid Technique [47]

This is a machine-generated summary of:

Ilango, R.; Vengadachalam, N.; Seethalakshmi, V. Subha: Charging demand based on the interaction among electric vehicles and renewable energy sources using hybrid technique [47].

Published in: Clean Technologies and Environmental Policy (2022).

Link to original: https://doi.org/10.1007/s10098-022-02334-w.

If you want to cite the papers, please refer to the original.

For technical reasons we could not place the page where the original quote is coming from.

Abstract-Summary

"An efficient hybrid technique is proposed for optimal allocation of electric vehicle charging spots in distribution network system."

"The main purpose of the proposed technique is "to decrease the real and reactive power loss through the optimal allocation of the electric vehicles parking lot, and also decreases the cost." "

"The proposed technique is used to evaluate the reliability of the smart grid based on the state matrix incorporation of random parameters of the electric vehicle."

"The proposed technique helps to increase the calculation speed by detecting the desired count of electric vehicle states required."

Introduction

"Numerous methods have been proposed to assign EV charging stations such as comprehensive probability model, parking-based allocation system, and particle swarm optimization (PSO) along genetic algorithm (GA)."

"The results show that with the optimal placement of electric vehicles as parking, this process is less dependent not only on incentives, other than barriers for comfortable driving."

"Objectives and contribution (a) In the proposed work, an efficient hybrid technique is proposed for optimal allocation of electric vehicle charging spots in distribution network system."

"The proposed technique is implemented in the MATLAB/Simulink working platform and its performance is compared with existing techniques. (b) The novelty of the proposed work is "to obtain the overall optimal solution and the optimal solution is gathered from the search space of these operators." (c) The intention of the FBIAOA work is decreasing the loss of real and reactive power through the optimal allocation of the electric vehicles parking lot and lessens the cost."

Literature Review

"Several research works were presented based on the optimum allocation of EV charge plug using different methods and aspects."

"Andrenacci and others [48] have introduced the demand side method for the optimum exploitation of electric vehicle charge stations within metropolitan zones."

"Song and others [49] have presented the convexity method for the programming of electric vehicle charge stations that diminishes the entire charging rate with energy cost of substation through time distance in substation."

"Charge facilities through dissimilar valued charge power may meet the charge demands of various electric vehicle owners and at the same time affect the three-dimensional as well as temporal distribution of electric vehicle charge demands, thereby challenging rationality employs the economics of EVCS assignment system."

"This occurrence creates three-dimensional with temporal distribution of EVCS additional complex; therefore, the huge significance employs to catch an effectual method as EVCS optimum development that encompasses multiple types of charging facilities."

Overview of Electric Vehicle Charging Based on the Mathematical Modeling of Renewable Energy Sources

"The first step is to design the electric vehicle requirement in charging station."

"This is necessary to measure the capacity of electric vehicles, including the charge level of the battery."

"The other random number in the E parameter is proposed for every electric vehicle coming to the charge station to obtain the battery charge status."

"The count of chargers that provide rapid charging power and power, m, the overall demand of electric vehicle station is: Operators of EV regulate the nature of its charge depending on the electricity cost to reduce the charging cost."

"Time is essential in fast dc charge station (480 V DC or above) that completely charging electric vehicle is lesser while using ac category."

"It proceeds an hour to charge 0 to 80% of electric vehicles filled battery volume."

Mathematical Modeling of Multi-Objective Functions

"The mathematical simulation of multiple objective function is presented as follows."

"For optimize charging and purchasing plan of EV is the main aim of the Ds, and it is utilized to enhance its charging performance in a comprehensive market (Chou and Nguyen [50])."

Proposed Approach of Forensic-Based Investigation Algorithm (FBI) and Archimedes Optimization Algorithm (AOA)

"If an object I is not present along the global optimum and the accelerating value is higher in the meaning of an object, it is exploration phase or the exploitation phase."

"Step 4: Perform cylindrical investigation and pursuit process Determine the investigation process based on the fitness function."

"Step 5: Analysis of location Analysis the location of the population based on the following equation: Step 6: Update the worst location The worst location is updated by the following equation: Step 7: Identify the target location To minimize the cost function of device, determine the target location of the system which is described as: Step 8: Update the location to head After receiving the optimum location of the system report back to head, upgrade the location by, Step 9: Update the location to head By using the above investigation, find the optimal solution."

Simulation Results with Discussion

"The proposed FBIAOA method flows as of 1 to 0.4$ at 1 to 5 h of time period, and it increases up to 1.2$ at 12 h of time period and it decreases up to 0.5$ at 20 h of time period, and then it increases up to 1.4$. The existing S2NA-CGO method flows as of 0 to 0.5$ at 5 h of time period, and at 10 h of time period, the existing method increases up to 1.4$, and at 24$ of time period, the existing method flows as of 1.4$. A subplot (b) the cost of FBIAOA and CHA is presented."

Conclusion

"An efficient hybrid technique is proposed for optimal allocation of electric vehicle charging spots in distribution network system."

"The main purpose of the FBIAOA technique is to decrease the loss of real and reactive power via the optimal allocation of the electric vehicles parking lot, and also decrease the cost."

"The proposed technique is used to evaluate the reliability of the smart grid based on the state matrix incorporation of random parameters of the electric vehicle."

"The proposed technique helps to increase the calculation speed by detecting the desired count of electric vehicle states required."

"The number of iteration of 100, 250, 500, and 1000 using proposed technique has the simulation time of 14.8, 29.2, 69.9, and 77.1 s. Optimal allocation of EV-CS based on distribution system performance and optimal/smart charging feature of EVs can also system performance significantly and considered as future scope of this research."

10.2.9 Determination of Maximum Additional Load for EV Charging Station Considering Practical Security Limits [51]

This is a machine-generated summary of:

Mondal, Sayak; Acharjee, Parimal; Bhattacharya, Aniruddha: Determination of maximum additional load for EV charging station considering practical security limits [51].

Published in: Journal of Engineering and Applied Science (2023).

Link to original: https://doi.org/10.1186/s44147-023-00204-w.

If you want to cite the papers, please refer to the original.

For technical reasons we could not place the page where the original quote is coming from.

Abstract-Summary

"Although a single EV has a negligible impact on the system, the combined effect of charging an EV cluster may violate the voltage security constraints of the RDS."

"Conventional analytical methods were used to calculate the maximum loadability of the RDS node, where voltage magnitudes become less than the critical voltage limit."

"The appropriate capacity of the EVCS can only be determined if the voltage security limit of each node of the RDS follows its critical limit."

"In this paper, a modified forward–backward sweep (MFBS) algorithm is formulated to find the maximum additional load (MAL) of each node of the RDS considering non-unity power factor during the EV charging process."

"This algorithm may help the EVCS to determine its capacity or optimal number of charging ports directly during installation at an optimal location of RDS."

"For validation, the 33-bus and 69-bus RDSs have been used, and it can be seen that the EVCSs have been successfully installed in the optimal location of the test systems mentioned above without violating the voltage security limit of each node of the RDS."

Introduction

"Considering these literatures along with its limitation, the scope of work is identified as (i) the authors must consider both active and reactive power together for doing the analysis related to the EV charging applications. (ii) As the voltage security limit is not analysed properly to find the loadability limit of each nodes of a RDS, the maximum loadability limit of a node should be determined in such a way that all the bus voltages of RDS must remain above the critical voltage limit. (iii) There should be a new modified index, which will not allow the system to go beyond the threshold of voltage limits."

"As the charging of an EV with optimum capacity is a challenging task, hence it is very important to identify the extra load of a node for which voltage magnitude of any node of that system should not be less than the critical voltage (V_{cri}) for the safe and secure operation of the power system."

Methods

"Based on the analysis as shown in [52], it can be revealed that forward–backward sweep (FBS) method is one of the most promising and efficient methods for load flow analysis in radial distribution system (RDS)."

"This method requires backward sweep for branch current calculation and forward sweep for determining the voltage."

"Power flow in each system branch, power injected by each generating source, voltage magnitude and phase angle at each system bus, and system losses are all determined using load flow (LF) analysis."

"Using FBS method first the additional extra active power (MAL) is calculated."

"To find the maximum allowable extra load for a node, a new additional factor MAL is added for increasing the load to a node in the MFBS load flow method."

"The update of system voltage will lead to updating the active power and reactive power at m-node in modified forward sweep."

Results and Discussion

"Considering this philosophy and based on the higher value of MAL, 8 EVCS at node 2, 3, 19, 20, 21, 22, 23, and 24 are chosen for 33-bus RDS."

"MAL is determined for all buses of 69-bus RDS and based on the higher value of MAL, 8 nodes are identified."

"For 33-bus RDS the critical voltage is obtained at node 18 after applying MAL on node 2."

"After calculating MAL for the various nodes of 33-bus RDS, nodes having higher MAL are decided as the optimal location of the EVCS."

"The maximum allowable load for the node 2 of the EVCS is 4776 kW which is higher than the MAL value of node 2 of RDS."

"In the case of 69-bus RDS, the EVCS connected node and other node's voltages exceed the critical voltage limit of 0.9 p.u."

Conclusions

"The existing conventional non-elastic load for the nodes of 69-bus is higher as compare to the 33-bus RDS."

"The capacity of the EVCS is higher at 33-bus RDS."

"It is obtained that the average loadability limit of each EVCS at 33-bus and 69-bus RDS are 3058 kW and 1272 kW respectively, whereas the total capacity of EVCS at 33-bus and 69-bus RDS are 24,464 kW and 10,176 kW. As the maximum loadability limit of 33-bus is high, hence more number of EV batteries can be connected for charging in case of 33-bus RDS compared to 69-bus RDS."

"It can be said that the proposed algorithm is efficient enough for maintaining the node voltages during EV charging process for both higher and lower bus RDS."

Bibliography

1. Morfeldt J, Johansson DJA (2022) Impacts of shared mobility on vehicle lifetimes and on the carbon footprint of electric vehicles. Nat Commun. https://doi.org/10.1038/s41467-022-33666-2
2. Nordelöf A, Messagie M, Tillman AM, Ljunggren Söderman M, van Mierlo J (2014) Environmental impacts of hybrid, plug-in hybrid, and battery electric vehicles—what can we learn from life cycle assessment? Int J Life Cycle Assess 19:1866–1890. https://doi.org/10.1007/s11367-014-0788-0
3. Hawkins TR, Singh B, Majeau-Bettez G, Strømman AH (2013) Comparative environmental life cycle assessment of conventional and electric vehicles. J Ind Ecol 17(1):53–64. https://doi.org/10.1111/J.1530-9290.2012.00532.X
4. IEA. World Energy Outlook 2020 (2020). www.iea.org/weo
5. Nunes A, Woodley L, Rossetti P (2022) Re-thinking procurement incentives for electric vehicles to achieve net-zero emissions. Nat Sustain. https://doi.org/10.1038/s41893-022-00862-3
6. Burlig F, Bushnell J, Rapson D, Wolfram C (2021) Low energy: estimating electric vehicle electricity use. AEA Pap Proc 111:430–435
7. Fjørtoft TO, Pilskog GM (2019) A quarter of the richest households have an electric car (Statistics Norway, 2019). www.ssb.no/transport-og-reiseliv/artikler-og-publikasjoner/kvart-sjette-av-dei-rikaste-hushalda-har-elbil
8. Xing J, Leard B, Li S (2021) What does an electric vehicle replace? J Environ Econ Manag 107:102432
9. Insights into Future Mobility (MIT Energy Initiative, 2019). http://energy.mit.edu/insightsintofuturemobility
10. Ojha R, Agarwal A (2023) Implications of circular production and consumption of electric vehicle batteries on resource sustainability: a system dynamics perspective. Environ Dev Sustain. https://doi.org/10.1007/s10668-023-03279-w

11. Zhao H, Guo S, Zhao H (2018) Comprehensive benefit evaluation of eco-industrial parks by employing the best-worst method based on circular economy and sustainability. Environ Dev Sustain 20(3):1229–1253
12. Sun C, Li Z, Wang X, Li C (2019) Technology development of electric vehicles: a review. Energies 13(1):90
13. Alessia A, Alessandro B, Maria VG, Carlos VA, Francesca B (2021) Challenges for sustainable lithium supply: a critical review. J Clean Prod 300:126954
14. Aichberger C, Jungmeier G (2020) Environmental life cycle impacts of automotive batteries based on a literature review. Energies 13:6345
15. Gautam A, Shankar R, Vrat P (2022) Managing end-of-life solar photovoltaic e-waste in India: a circular economy approach. J Bus Res 142:287–300
16. Guarnieri P, Cerqueira-Strei JA, Batista LC (2020) Reverse logistics and the sectoral agreement of packaging industry in Brazil towards a transition to circular economy. Resour Conserv Recycl 153:104541
17. Jabbour CJC, de Sousa Jabbour ABL, Sarkis J, GodinhoFilho M (2019) Unlocking the circular economy through new business models based on large-scale data: an integrative framework and research agenda. Technol Forecast Soc Change 144:546–552
18. Wendler T (2019) About the relationship between green technology and material usage. Environ Resour Econ 74:1383–1423
19. Rosendahl KE, Rubiano DR (2019) How Effective is lithium recycling as a remedy for resource scarcity? Environ Resour Econ 74:985–1010
20. Dixit SK, Singh AK (2022) Predicting Electric Vehicle (EV) buyers in India: a machine learning approach. Rev Socionetwork Strateg. https://doi.org/10.1007/s12626-022-00109-9
21. Kumar R, Jha A, Damodaran A, Bangwal D, Dwivedi A (2021) Addressing the challenges to electric vehicle adoption via sharing economy: an Indian perspective. Manag Environ Qual Int J 32(1):82–99. https://doi.org/10.1108/MEQ-03-2020-0058
22. Janamala V, Kamal Kumar U, Pandraju TKS (2021) Future search algorithm for optimal integration of distributed generation and electric vehicle fleets in radial distribution networks considering techno-environmental aspects. SN Appl Sci. https://doi.org/10.1007/s42452-021-04466-y
23. Kowsalya MIA (2014) Optimal distributed generation and capacitor placement in power distribution networks for power loss minimization. In: Proceedings of international conference on advanced electrical engineering (ICAEE), pp 1–6. https://doi.org/10.1109/ICAEE.2014.6838519
24. Rajeswaran S, Nagappan K (2016) Optimum simultaneous allocation of renewable energy DG and capacitor banks in radial distribution network. Circ Syst 7(11):3556–3564. https://doi.org/10.4236/cs.2016.711302
25. Ali ES, Abd-Elazim SM, Abdelaziz AY (2018) Optimal allocation and sizing of renewable distributed generation using ant lion optimization algorithm. Electr Eng 100(1):99–109. https://doi.org/10.1007/s00202-016-0477-z
26. Truong KH, Nallagownden P, Elamvazuthi I, Vo DN (2020) A quasi-oppositional-chaotic symbiotic organisms search algorithm for optimal allocation of DG in radial distribution networks. Appl Soft Comput 88:106067. https://doi.org/10.1016/j.asoc.2020.106067
27. Truong KH, Nallagownden P, Elamvazuthi I et al (2020) An improved meta-heuristic method to maximize the penetration of distributed generation in radial distribution networks. Neural Comput Appl 32:10159–10181. https://doi.org/10.1007/s00521-019-04548-4
28. Nguyen TT, Nguyen TT, Nguyen NA, Duong TL (2020) A novel method based on coyote algorithm for simultaneous network reconfiguration and distribution generation placement. Ain Shams Eng J. https://doi.org/10.1016/j.asej.2020.06.005
29. Wu Y-K, Chang SM, Hu Y-L (2017) Literature review of power system blackouts. Energy Procedia 141:428–431. https://doi.org/10.1016/j.egypro.2017.11.055
30. Singhal R, Patel CR (2022) Prioritisation of barriers in the implementation of electric mobility in Indian context using fuzzy analytical hierarchical process. J Inst Eng (India): Ser A. https://doi.org/10.1007/s40030-022-00676-8

31. Larimian T, Zarabadi ZS, Sadeghi A (2013) Developing a fuzzy AHP model to evaluate environmental sustainability from the perspective of Secured by design scheme: a case study. Sustain Cities Soc 7:25–36. https://doi.org/10.1016/j.scs.2012.10.001
32. Awang A, Ghani AA, Abdullah L, Ahmad MF (2017) Fuzzy analytic hierarchy process (FAHP) with cosine consistency index for coastal erosion problem: a case study of setiu wetlands. J Comput Sci Comput Math 7(4):107–118
33. Saad SM, Kunhu N, Mohamed AM (2016) A fuzzy-AHP multi-criteria decision making model for procurement process. Int J Logist Syst Manag 23(1):1–24
34. El-Din HK, El Munim HE, Mahdi H (2019) Decision-making in fuzzy environment: a survey (IntechOpen). Cairo. https://doi.org/10.5772/intechopen.88736
35. Arora SC, Sharma M, Singh VK (2022) Using diffusion of innovation framework with attitudinal factor to predict the future of mobility in the Indian market. Environ Sci Pollut Res. https://doi.org/10.1007/s11356-022-23149-8
36. Pradeep VH, Amshala VT, Kadali BR (2021) Does perceived technology and knowledge of maintenance influence purchase intention of BEVs. Transp Res Part D: Transp Environ 93:102759. https://doi.org/10.1016/j.trd.2021.102759
37. Xu G, Wang S, Li J, Zhao D (2020) Moving towards sustainable purchase behavior: examining the determinants of consumers' intentions to adopt electric vehicles. Environ Sci Pollut Res 27(18):22535–22546
38. Kahn ME (2007) Do greens drive Hummers or hybrids? Environmental ideology as a determinant of consumer choice. J Environ Econ Manag 54(2):129–145. https://doi.org/10.1016/j.jeem.2007.05.001
39. Degirmenci K, Breitner MH (2017) Consumer purchase intentions for electric vehicles: is green more important than price and range? Transp Res Part D: Transp Environ 51:250–260
40. Rezvani Z, Jansson J, Bodin J (2015) Advances in consumer electric vehicle adoption research: a review and research agenda. Transp Res Part D Transp Environ 34:122–136. https://doi.org/10.1016/j.trd.2014.10.010
41. Sajjad A, Asmi F, Chu J, Anwar MA (2020) Environmental concerns and switching toward electric vehicles: geographic and institutional perspectives. Environ Sci Pollut Res 27(32):39774–39785. https://doi.org/10.1016/j.enpol.2012.04.055
42. Jaiswal D, Kant R (2018) Green purchasing behaviour: a conceptual framework and empirical investigation of Indian consumers. J Retail Consum Serv 41:60–69. https://doi.org/10.1016/j.jretconser.2017.11.008
43. Shalender K, Sharma N (2021) Using extended theory of planned behaviour (TPB) to predict adoption intention of electric vehicles in India. Environ Dev Sustain 23(1):665–681
44. Anderson JC, Gerbing GW (1988) Structural equation modeling in practice: a review and recommended two-step approach. Psychol Bull 103:411–423
45. Hair JF, Anderson RE, Tatham RL, Black WC (1998) Multivariate data analysis with readings. Prentice-Hall, Englewood Cliffs
46. Li W, Long R, Chen H, Geng J (2017) Household factors and adopting intention of battery electric vehicles: a multi-group structural equation model analysis among consumers in Jiangsu Province. China Nat Hazards 87(2):945–960. https://doi.org/10.1007/s11069-017-2803-9
47. Ilango R, Vengadachalam N, Seethalakshmi V (2022) Subha Charging demand based on the interaction among electric vehicles and renewable energy sources using hybrid technique. Clean Technol Environ Policy. https://doi.org/10.1007/s10098-022-02334-w
48. Andrenacci N, Ragona R, Valenti G (2016) A demand-side approach to optimal electric vehicle charging stations in metropolitan areas. Appl Energy 182:39–46. https://doi.org/10.1016/j.apenergy.2016.07.137
49. Song Y, Zheng Y, Hill DJ (2016) Optimal scheduling for EV charging stations in distribution networks: a convexified model. IEEE Trans Power Syst 32(2):1574–1575
50. Chou JS, Nguyen NM (2020) FBI inspired meta-optimization. Appl Soft Comput 93:106339
51. Mondal S, Acharjee P, Bhattacharya A (2023) Determination of maximum additional load for EV charging station considering practical security limits. J Eng Appl Sci. https://doi.org/10.1186/s44147-023-00204-w

52. Shirmohammadi D, Hong HW, Semlyen A, Luo GX (1988) A compensation-based power flow method for weakly meshed distribution and transmission networks. IEEE Trans Power Syst 3(2):753–762. https://doi.org/10.1109/59.192932
53. Pehl M et al (2017) Understanding future emissions from low-carbon power systems by integration of life-cycle assessment and integrated energy modelling. Nat Energy 2:939–945
54. European Commission (2018) In-depth analysis in support of the Commission communication COM (2018) 773—a clean planet for all—a European long-term strategic vision for a prosperous, modern, competitive and climate neutral economy. https://ec.europa.eu/clima/policies/strategies/2050_en
55. Swedish Energy Agency (2021) Scenarier över Sveriges energisystem 2020 (Scenarios for Sweden's energy system 2020) ER 2021:6. https://energimyndigheten.a-w2m.se/Home.mvc?ResourceId=185971

The manufacturer's authorised representative in the EU is Springer Nature Customer Service Centre GmbH, Europaplatz 3, 69115 Heidelberg, Germany. If you have any concerns regarding our products, please contact ProductSafety@springernature.com

Printed and bound by CPI Group (UK) Ltd, Croydon, CR0 4YY
01/10/2026
02225910-0010